Sociology and Environmental Sustainability

Sociology and Environmental Sustainability

A Study of Fragile Ecology, Health Risks and Population Pangs

Rani Mehta

Rawat Publications

Jaipur • New Delhi • Bangalore • Hyderabad • Guwahati • Kolkata

ISBN 978-81-316-0630-8

Published by
Prem Rawat for **Rawat Publications**
Satyam Apts, Sector 3, Jawahar Nagar, Jaipur 302 004 (India)
Phone: 0141 265 1748 / 7006 Fax: 0141 265 1748
E-mail: info@rawatbooks.com
Website: www.rawatbooks.com

New Delhi Office
4858/24, Ansari Road, Daryaganj, New Delhi 110 002
Phone: 011 2326 3290

Also at *Bangalore, Hyderabad, Guwahati* and *Kolkata*

Typeset by Rawat Computers, Jaipur
Printed at Chaman Enterprises, New Delhi

Contents

Preface

The planet Earth is unique because it provides life on it to micro-organisms, plants and animals, including Man. It is observed that a continuous interaction takes place between organic and inorganic elements existing in the environment. This has helped Man to evolve himself through the natural selection process, adaptation and genetic mutations. As a consequence of that, he has developed an ecology inclusive of both natural and biological elements, encompassing the total environment. No doubt, the environment over several millennia has been subjected to certain intrinsic and extrinsic changes through technological innovations and discoveries engineered by Man, for improvement in his lifestyle. Yet, these changes along with better economic development have also resulted in affecting life of people adversely. Because of growth of living habitats due to over-population, slums, water and sanitation crises, flow of industrial and chemical effluents, inadequate sewerage for solid waste disposal, air pollution due to emission of deadly gases by factories and automobiles of various kinds, etc., life is not only becoming difficult but miserable as well. The consequences of economic growth over the years have been observed to be a matter of serious concern both for the developed as well as the developing nations. In our context, these have assumed dangerous proportions because of faster population growth, resource crunch, non-development of technology for cleaner environment, generation of power through the burning of coal and oil, which are receding in quantity and non-availability and inadequacy of renewable energy technology through sun, water and wind.

These factors forced us to take risks with nuclear energy generation in collaboration with the developed nations. All these issues are acquiring crises situations as the environment has deteriorated so much that one cannot lead a minimum decent life. In fact, the resultant effect of economic development has led to a fragile ecology created by Man for his affluent lifestyle. Thus, it has become a boon for some but a bane for the majority.

This situation has to be reversed by the genius of Man to evolve technology or methods which should be environment-friendly so as to provide environmental sustainability to our future generations, otherwise we are likely to get affected badly. Instead of relying heavily on the techno-economic system, to which we have enslaved ourselves, we need to look for alternative environmental sustainable efforts based on folk wisdom, indigenous knowledge and cultural practices of pluralized groups to preserve and protect our environment by evolving suitable or intermediate technological innovations. Further, there is a need to have community participation and involvement of people to provide sustainable development both in rural and urban areas. This is what the need of the hour is. This should be seriously pursued for a healthy environmental living and for maintaining its sustainability.

In my endeavour to take up this task, I was inspired by a course on Environmental Crisis and Sustainable Development which I happened to teach to the post-graduate and research students of Sociology at Panjab University, Chandigarh for several years. Further, I was encouraged by Chairpersons of the Department (Professor M.K. Teja, Professor Sherry Sabbarwal, Professor Rajesh Gill and Professsor Kiran Preet Kaur) to take up teaching of this course and they deserve all my appreciation and thanks. I shall also like to mention the encouragement I got from my parents (Professor S.R. Mehta and Mrs Saroj Mehta), who despite many odds encountered by me, supported me to complete this task. The feedback I got from my students while teaching this course clarified my doubts and made my concepts clear. They also deserve my thanks. I also take this opportunity to extend my thanks to Mr Pranit Rawat of Rawat Publications, Jaipur for expediting the publication of the book.

Last but not least, I am thankful to Mr Hari Ram Yadav for the arduous work of typing of the manuscript and bringing it in the present form.

Rani Mehta
Centre for Advanced Study in Sociology
Punjab University, Chandigarh

1

Introduction

In the celestial planetary system, the planet earth concerns us the most because as humans, we live on it. Notwithstanding its interacting influences with other planets and luminaries such as sun and moon, from which it gets energy and vibration, it is observed that nearly 29 per cent of land area of the planet provides the habitat for plants and animals, while about 71 per cent is surrounded by water.

Planet earth is unique in its characteristic in the sense that it has provided life on it to micro-organisms, plants and animals, including man. Existence of life on other planets is in the stage of exploration as it is surrounded by mysteries, uncertainties and speculative explanations that deny scientific validity about it.

The existence and perpetuation of life on planet earth have been scientifically explained by the evolved Man through Darwin's law of evolution, Mendelian's principles of inheritance and selection of genes and Newton's laws of thermodynamics denoting energy generation and dissipation, necessary for organic life. Man has evolved himself in interaction with micro-organisms, plants and animals in an environment surrounding him and fellow human beings. This environment was rough and unfriendly, abetted by hostile and ferocious animals, poisonous reptiles and harmful plants. He, through his understanding and reasoning capabilities, ultimately succeeded in harnessing such an environment to make his biotic (biological) life and its perpetuation possible and convert the physical or abiotic elements surrounding him – that is, space, land forms, water bodies, climate, soils, rocks, minerals, etc. – to his utmost use and advantage.

There is, no doubt, a continuing interaction taking place between the organic and inorganic elements existing in the environment and through the natural selection processes, adaptation and genetic mutations, etc., the plant and animal life, including that of man and other micro-organisms, got evolved in correspondence with the Darwinian law of 'survival of the fittest'. In this process, man developed an ecology inclusive of both natural and biological elements encompassing the total environment, with a view to provide a dynamic equilibrium between the organic and inorganic elements on earth. The relationship between biotic and abiotic elements on the planet needs to be explained by conceptualizing both the environment and ecology.

Environment and Ecology

How do we differentiate between environment and ecology? Environment has a wider connotation and it needs to be understood in the context in which it is employed. Broadly, it includes the surroundings or contexts within which humans, animals or objects exist or act. In a specific term, it may be referred to as the sum of outside influences on the organism in the context of learning and gaining experience from the surroundings by humans. However, we need to differentiate it from the inheritance potential of man, which also influences his or her development and behaviour in the context of nature-nurture debate. On the other hand, ecology in a broad sense has been referred to as the study of interactive relationship between living things and their environment. More specifically, it has been conceived as a branch of biology which deals with habits, modes of life and relations of living organisms with their surroundings. The term became popular in the 1980s when the environmentalists got concerned with the fragility of Earth as a living system. It was recognized that the natural systems which were evolved over millennia were getting threatened by the technological advancements initiated by the industrial revolution and the consequent population explosion. This became visible on planet earth due to extinction of many species of plants and animals, depletion of ozone layer, global warming and changes in weather pattern and pollution of large areas of land, water and air, disturbing the natural balance of many smaller systems.

Over the years, as a result of intensification of agriculture, mining, industrial pollution and physical overcrowding, the available physical and economic resources in the ecosystem have either run down or become degraded to the extent that a situation has arisen where for the minimum decent living of human beings, we urgently need to pay careful attention to the degradation of physical and social environment. Notwithstanding the acceleration of such processes in the industrial and post-industrial societies, it has been recognized by Harris (1978) that this process has been in existence in human societies for a longer period. His contention is that the Stone Age people might have lived far happier and healthier lives than those who came after them. In fact, he is of the opinion that many of the cultural, political and socio-economic transformations encountered by societies in the modern or historical period may be explained as consequences of environmental pressures, such as the need for settled agriculture, subordination of women, a necessity for state direction and prohibition of meat eating in some cultures. However, such claims may appear controversial but it is a recognized fact that environmental depletion is not a recent problem and has serious implications for humans.

Human Ecology: Natural vis-à-vis Social

The tradition of human ecology gave rise to the birth of environmental sociology. The 'New Environmental Paradigm' as suggested by Riley Dunlap and his associates tries to redefine the relationship between human societies and their natural environments. His notion of 'web of life' underscores the interconnectedness, social and natural processes and factors which simultaneously act as an instrument for providing a critique on the borderline notions such as the New Ecological Paradigm (NEP) as against the Human Exemptionalist Paradigm (HEP) that dominated the social sciences (Dunlap and Catton, 1979, 1994). The latter paradigm not only defined the environment as something 'out there' but also showed no concern about its possible impacts on human society. This debate, no doubt, contributed significantly to the establishment of a new field of study within the social sciences, but it turned out to be a dead-end mainly because the boundaries between the social and the natural worlds were not settled in the human ecology tradition of the Chicago school, which had emphasized that there was a realm within societies that was shaped primarily by non-social or sub-social factors, that is the biotic community (Nelissen, 1972; Spaargaren et al., 2000).

It is argued that within the biotic sub-sphere of society, there are certain phenomena and facts that need to be explained in the context of biological and natural sciences. Further, it is opined that the biotic community is counterposed to society or culture and both these realms are likely to have their own different logics or dynamics. It is also emphasized that the non-social sphere of biotic community is supposed to be governed by ecological laws or dynamics, which need to be explained through concepts such as survival of the fittest, adaptation, invasion, over-shoot and so on. If the social system is not affected by human-made policies and politics, then one may expect same patterns to occur as a result of these bio-natural mechanisms (Spaargaren et al., 2000).

If the basics of human ecology are valid, then, one may infer that crossing the society-nature border within the human ecology tradition would take us 'beyond the social' in a specific manner. It is further argued by Gert Spaargaren et al. (2000) that right from Chicago school up to Peter Dickens' (1992) book *Society and Nature*, we need to understand some social facts or behavioural patterns and explain these by laws or regularities that exist across different human and non-human species. For example, behavioural practices in the field of child-raising, dating and mating or defending a territory or hierarchical order are considered to be governed and thus need to be explained by making reference to certain basic laws that may take us beyond the social realm and logics.

Environment and Technology

No doubt, technology whether primitive or advanced has been instrumental in exploiting the environmental resources to give impetus to the development of mankind. Technology is evolved on the principles of science through systematized studies of various natural phenomena such as physical and biological in an environmental system by the input of solar energy to generate organic matter. Organic life is the centre stage of the environment, influenced by processes both within and outside the biospheric environment, generated by endo-genetic and exo-genetic forces. In analysing technology for the exploitation of environmental resources to the advantage of mankind, there is a different emphasis laid on the nature and impact of technology on society. In certain quarters, there is thrust laid on the dominant material or source of energy in use (animate inanimate) while in other quarters, there is a focus on the impact of particular techniques such as mechanical, intellectual, social and communicative. In essence, the idea is to have practical application of knowledge and use of techniques in productive activities. However, in sociological parlance, we need to treat technology as a social product, which includes the human artifacts such as tools and machines and the knowledge as well as ideas involved in different productive activities which may not always depend upon science as a driving force, as was witnessed in the relatively simple forms of mechanization in the early industrial revolution. However, recent advances in energy production and information technology have evolved many kinds of innovations mainly derived from organized science.

The role of technology in social change has been of great concern in sociology right from the Marxist analysis of the forces and relations of production to theories of industrialization, modernization of post-industrial and information societies. These later theories were mainly developed in 1960s and were mainly derived upon neo-evolutionary assumptions. In fact, technology has always been accorded a key determining role in shaping the social structure of advanced industrial societies.

Technology not only changes the physical environment but it also has implications for socio-cultural system. Ideas are central in the advancement and development of technology. Man, while making an attempt to exploit nature to his advantage and with the expansion of population, always believed in progress and showed a concern for posterity within a linear concept of time (Black, 1970).

This attitude of man towards natural environment was based on Hebrew theology and its Christian interpretation, which endorses that 'Be faithful and multiply, and replenish the Earth and subdue it, and have dominion over the fish of the sea and over the fowl of the air and over every living thing that moved upon the Earth' (ibid.). However, there has been a potential conflict between the extreme exploitation of nature and the well-being of man. This may get mediated by a sense of responsibility to the family in terms of extending it to future generations and to God – a duty of stewardship. Rationalizing the use of resources and allocation of resources to different groups became basic to the development of capitalism through unequal and unlimited accumulation. But, in the context of a subsistence economy in the early stages of food-gathering and hunting as well as domestication of plants and animals, the family income was generated by its own efforts to contain self-sufficiency and the market value of the product did not arise. However, in an industrial society, Max Weber attributed the process of rationalization to the development of large-scale industrial and development organization for a modern world. Its influence can also be traced through the development of science and technology. It is suggested that a perspective which sets man apart from nature and urges him to dominate it supports the idea of scientific inquiry (Harvey and Hallett, 1977). In this context, Schwartz (1971) visualized a transition through science from a subjective to an objective, a human and in the end an inhuman approach. In fact, he argues that man himself becomes an object in a mechanical world, to be studied and controlled as for him 'science has become a secular religion, technology its temple, efficiency its dogma'.

It has been observed that the basis of technology, no doubt, is creativity and innovation on the part of mankind. However, there are limitations of technology to harness or exploit environment in totality. In fact, technology conditions environment but it also gets conditioned by certain unintended consequences of the rough and tough environment, which may result in natural hazards such as tsunami-like conditions, famine and drought due to scanty rainfall and uncontrollable generation of gases and spread of air and water pollution (in the absence of non-availability of clean technology in certain parts of the planet earth). These may cause health risks and prove counter-productive to the development of mankind. It is in this context that Schumacher (1973), through his intermediate technology development group, advocated alternative technologies (which have been expressed in a variety of terms such as intermediate, appropriate or low-impact technology for development of human societies, especially in the context of developing nations having limited technological advancements). His argument is that advanced technology is capital rather than labour-intensive so it is not suitable to most situations in developing countries, where simple, rugged equipment is readily available and can be used and maintained by local resources.

But, over the years, we have witnessed the growth and development of an industrial society through the intensive use of inanimate energy resources for mass production of goods and services and the ability to tap massive energy reserves. The need for products is unlimited because human wants are insatiable and this is supported by a psychological need for self-acquisition and accumulation. But, recent advances in technological innovations have made the human relationships, institutions and values more complex than before as we witness a greater hierarchical pattern of stratification of societies as the users of technology tend to be those who can afford to avail of such products and services. For example, the main three latest technologies such as biotechnology, nanotechnology, communication or information technology have been evolved to enhance human capacities for development of mankind, but these may tend to make industrial or post-industrial societies more divisive in stratified terms rather than upholding the value of equitable distribution of resources and encouraging equal accessibility to them.

Environment and Economic Development

Survival is the uppermost requirement for man. In the early years, right from cave man through food-gathering and hunting stage to settled agricultural-based village life after the invention of fire, harnessing of water and domestication of plants and animals, men in order to eke out for their existence had to work together as a group for meeting their basic needs like food, shelter, clothing and sex gratification. The evolution of group life through the mating process of opposite sex and formation of families increased the interaction of men, women and children with greater intensity to the immediate surrounding or environment for making social life a reality. In fact, all economic phenomena as pursuits for providing a minimum decent and comfortable life are basically social in nature and together constitute the various processes for evolving values and norms that govern social relationships through an authority structure or power system to regulate the behaviour patterns of individuals, groups, communities and societies.

From pastoral or subsistence agricultural, pre-industrial and post-industrial growth and development of communities and societies, man has employed his genius, intelligence, ideas and values to exploit his surroundings or environment to his maximum advantage. However, economic development in relation to environmental explanation became a necessity for meeting the basic and other welfare needs of mankind. The first and the foremost factor for securing food for the increasing population was taken care of by using animate or inanimate energy by raising the food production of various crops that were possible to be grown within a particular ecological setting.

The basic factors involved in economic development have been land, human capital and non-human capital, usage of energy or technological innovations and entrepreneurship to extract, manufacture, use and provide the disposal of products. Over the last 200 years or so, there has been a tremendous increase in both the quantity and the complexity of man's demands for material and energy, much beyond mere survival level. In fact, these increases are a part of the process of development witnessed during the last more than 200 years of history and in modern economics, it is referred to as developmental planning in the context of both developed and developing countries (Harvey and Hallett, 1977).

It has also been reported by these scholars that the extraction-manufacture and use-disposable process has brought in a new path in the general schema of the natural cycle of a particular material, than the natural rate of a cycle involving different chemical combinations having varied concentrations in a different geographical location. It has no doubt called for the resilience and adaptability of the ecosystem, for example, in the context of direct human use of food which is a part of the natural living system, there appears less disturbance than the use of iron in a manufacturing process, as these are two different forms of resources. Resources may be classified as renewable (manageable and directed with regard to concentration in space, rate of extraction such as solar energy, food crops and water) and non-renewable (mineral resources and other metallic core concentrations which provide replacement materials over millions of years and cannot be recycled). To use the non-renewable resources in industrial processes, we need expanded energy, which is available in greater concentration than even in solar energy or human muscular energy and draught animals employed in the mining industry during the 16th century as a short-cut method across a natural cyclic path of processing of materials. Till the industrial revolution, it is observed that the scale of such operations was small but the growth in demand for metals has been so rapid since the beginning of Second World War that it exceeded all the production in the whole history of the world upto that time. This could be made possible by generating energy in the form of concentrated fuels, so it may be worthwhile to highlight briefly the availability of energy resources and their harnessing (ibid.).

We need to make a distinction between energy and fuels because 'all fuels are sources of energy but not all sources of energy are fuels'. Fuels constitute some kind of materials that can be transformed to yield energy through suitable processes to rearrange the electronic structure of the atoms (or the structure of the nucleus in case of nuclear fuels) to release energy needed to maintain the structure. For example, oil is a chemical fuel which releases large quantities of energy after the rearrangement of electronic structure during burning. Fuels, at present, are the major sources of energy for the world. Broadly, these fall into three main groups. The first category is that of foodstuff, that is, energy sources for man and other animals, animal waste especially cow-dung (which is an important domestic fuel in Asia) and wood. This category draws energy mainly from photosynthesis. In the second group, we have fossil fuels, coal and oil and the associated shales, tars and natural gas. Till recently, this group also derived energy by photosynthesis, but it can also be concentrated by the long-term geological processes resulting in higher energy content per unit of the mass. The third group is that of nuclear fuels, which derive their energy through much more violent processes that occur before or during the formation of the earth. These have an enormously greater energy yield and can be extracted by processes that are far more complex and dangerous than burning. Only in the case of first category we can have the rate of production and the rate of usage as comparable and thus, we can describe this group as renewable source of energy. On the other hand, it is estimated that oil takes some 250 million years in its formation and the present world rate of its use is nearly ten times every second. This difference in the two rates makes us infer that the fossil fuels in the second group are non-renewable energy sources.

Till the 17th century, man was dependent on muscle power (his own as well as that of draught animals) and burnt wood to meet his energy needs. However, by 1850, coal turned out to be an important world fuel and nearly a little more than a quarter of the total as compared to wood, which remained just under a half of the total, as the principal source of energy. However, by 1900, coal use was increasing but oil was still not a significant contributor. Fifty years later, oil had become as important as coal and its use started growing rapidly. In the 1960s, it was observed that nuclear fuels began to contribute and their use started growing very rapidly.

The changing world scenario for the use of energy clearly brings out that man's needs are now increasingly being met by non-renewable sources. This poses a serious question as to the length of time for which the world demand for energy can be met from a fixed stock, growing at a rate of about 5 per cent per annum (doubling in 14 years). The simple answer to this is that it cannot be more than a few human lifespans unless there is some major change brought in the pattern of fuel use. More specifically, for each fuel source, we need to work out the demand pattern and have reliable estimation of the actual quantities that might be available.

It has been observed that over the last eight years or so, oil produced per metre of exploratory drilling has gone down and we are finding difficulty in exploring new oil drillings and simultaneously, the oil reserves are getting depleted. It is reported that oil production reached a peak in 1990 but will fall off to a negligible value by 2080. In fact, it is suggested that nearly 80 per cent of all the oil produced will be during the 70 years from 1955 to 2025, which is a very small episode in man's time-scale activities. In the case of coal, it is worked out to be nearly a 300-year period, for 80 per cent of the total coal availability.

No doubt, these fossil fuels have great advantages over wood in terms of energy per unit mass and over nuclear fuels in terms of simplicity and safety, but these have remained dominant fuels for transportation and other much more convenient small-scale applications. However, oil is not only a fuel but it has also been utilized as a rare material for fertilizers and pesticides, for the plastics industry and for many synthetics, construction materials and medical products. Its increasing and well-established use, even when we are aware of its short life-time as a world resource, makes us ponder whether we should burn it or not and destroy its unique chemical structure or not.

As we are foreseeing a decline of fossil fuel energy sources in the future, there is an attraction and fascination for the generation of large quantities of energy from relatively small masses of nuclear fuels. For example, one gram of uranium 235 has the energy equivalent to 2.7 tonnes of coal for electricity generation at a working station, thereby showing an efficiency of about 33 per cent to liberate 1,000 megawatt of electricity by consuming only 3 kg per day. This process of energy generation on a large scale is complex and dangerous and may not be suitable for individual transport and domestic use, but would be useful for large-scale electrical generation (ibid.).

As the world population is increasing and energy needs are rising, we may be compelled to have less dependence upon fossil fuels, but the moot question is whether or not nuclear power is a safe replacement on a world scale. Solar energy is, no doubt, another alternative that may attract especially the nations bestowed with sunshine, but its utilization may depend on a decentralization of 'power' generation in contrast to the needs of nuclear power to match the diffused nature of the supply. We have evolved other energy conversion techniques such as nuclear fusion reactors, magneto-hydro dynamic generators and fuel cells at the early stages of development. Broadly, from a technological perspective, future sources may be divided into two groups: one which require high capital input, large-scale and complex technology, while the other which can be made simply and on the small scale, needing comparatively lesser capital. This will, however, depend upon the difficult decisions we may take about future power supplies, mainly constrained by social and political issues related to the degree of centralization and the use of social units (ibid.).

With the advancement of scientific knowledge and technological innovations, the nation-states which acquired fossil fuel technology in terms of large quantity of their natural resources could extract and exploit them to the greatest advantage of their populations. Through innovations, they could raise the standard of living of the people by providing them with electricity, clean water, housing, road networks, easy and fast means of communication and transportation, goods generated by extraction and manufacturing industries, food security through providing nutritious and rich protein diets, education, health care, and gadgets which are electrically generated and run. This could become possible because of the expansion of steel industry during the industrial revolution, which had its beginning in the 18th century but got accelerated during the 19th century. However, it got into peak after the two World Wars in the 20th century, especially after the Second World War. From 1950 onwards, European nations and the United States of America acquired technological sophistication to provide a decent and comfortable life to majority of their respective populations, but the fallout of these industrial and post-industrial societies due to overexploitation of resources and burning of fossil fuels caused a devastating damage to their immediate environments as well as in the general atmosphere of the planet through effects of global warming, toxicity, green house affects of gases, water and air pollution, soil degradation and salinity, desertification, extinction of plants and animals, deforestation, ozone depletion, floods and tsunami-like conditions, etc. All such phenomena occurring in different geographical regions of planet earth depending upon the extent and intensity of fossil fuel, have resulted in generating fragile ecological and regional imbalances across the globe and made the life of most of the people unfit for decent living. Further, it led to increasing health risks and the populations even in the developing nations, which were in the process of industrialization and urbanization, also became vulnerable to such effects which made their life miserable due to limited resources and non-availability of clean technology.

The major pollutants which have emerged causing damage to the environment are mercury (broken thermometers released 7,000 kg of mercury each year in Canada alone); agricultural fungicidal treatments and paints; lead affecting the human blood stream through food, air and water, etc., causing serious damage to the central nervous system; oil whether spilled over into sea waters or waste available from shore-based industries or as a result of incomplete combustion due to fallout of air-borne hydrocarbons, are not only toxic to marine life but also may cause respiratory ailments, etc. Even the pesticides which are highly soluble in oil may result in long-term disruption of food items and lead to serious illnesses and chronic diseases on wider areas covered under agricultural operations. The other pollutants which cause air pollution are a combination of smoke and fog which occurs in the atmosphere over a period of time due to its concentration as smog mainly emanating from tall chimneys, wind, burning of certain products causing health hazards of various kinds; nitrogen which gets washed out of the soil into rivers and lakes being used as a fertilizing agent in agricultural growth; waste heat caused by burning of water plants and animals; the synthetic organic pesticides, including insecticides and herbicides, and the disturbance of the stratosphere which shields the earth between 12 and 50 kilometres above it from harmful radiations of the sun. (The stratosphere provides an ozone layer to absorb heat of the sun and if it gets disturbed, the chances of ultra-violet radiation increase). The last but the most significant pollutant is nuclear radiation, which causes damage to human tissue. This has been introduced in the atmosphere through many new radioactive sources, especially the fallout of radioactive materials from the testing of nuclear weapons and the use of radiation, particularly X-rays, in medicine. Notwithstanding the damages caused to the environment, there is now a hue and cry being raised against industrialization on such a grand scale to make human life unsustainable for long on this planet. This needs to bring reconciliation between economic models of development and the concerns of ecologists for evolving an alternative strategy which should be open and experiential in nature to focus on sustainable development across many generations (Norton, 2003).

There has to be a limit to economic growth and development as has been witnessed in the developed nations due to mass production, mass consumption and distribution in an industrial society. This is largely characterized by the separation of production from consumption; the potentiality of tapping massive energy reserves; the division of labour into specialized occupations and the establishment of a well-defined organizational structure. This has serious consequences for developing nations because most of them have pursued a capitalist path of development but due to lack of adequate resources, technical manpower, inadequate research and development in industry, lack of skilled and entrepreneurial manpower, inadequate food production, imperfect democratic governance and an imperfect market economy, there is a rise in socio-economic disparities among different strata of populations, increasing population pressure on land resulting in low agricultural productivity, slow growth of industrial development, inequitable educational inputs, inequitable health and medical care, etc. It has also given rise to an elite class structure and marginalization of the majority of the population, having poor economic circumstances and social status. In view of the above, we need to examine the sociological aspects of the economic life which has emerged over the years, both in the context of developed and developing nations. Perhaps, in the context of developing nations, the sociological explanations have more relevance in terms of reorientation of economic policies while in the case of developed nations, there is a need to restrict unlimited growth and revise the economic theory of redistribution.

Environment and Sociology

Human actors and their institutions and values play a significant role in environmental issues affecting mankind. The wisdom of the sociological trinity such as Marx, Weber and Durkheim as classical theorists has thrown up ideas like false consciousness, traditional domination and organic solidarity, which have a bearing on almost all sociological notions, including contemporary postmodern ideas. Though it may be claimed by some influential sociological theorists such as Anthony Giddens that the context and assumptions of 19th century classical theorists are not relevant to the late 20th century and may need revision or dissolution in the context of emerging problems such as crisis of environment, yet, in conceiving environmental sociology, we cannot ignore the contributions of neo-ogburnianism of Fred Cottrell (1955), the functionalism of Walter Firey (1960), the attitude studies of the out-group recreation group and general environmental public opinion studies (Burch et al., 1972), the rural community sociology of natural resources (Burch and Field 1988), etc., which may be considered as proto-environmental sociology, mainly influenced by the classical traditional theories (Buttel, 2000).

From mid-1970s onwards, environmental sociology started gaining momentum and even non-sociologists such as Webb (1952), Wilkinson (1973) and Bookchin (1971) highlighted the lack of attention to the environment, especially bio-physical environment. A need was expressed to modify the western sociological theory to reflect the importance of biosphere and bio-physical forces in social structure and social life. It was emphasized further to include in environmental sociology the 'new human ecology' as a new ecological paradigm. In these approaches, there was some inherent hostility shown towards mainstream sociology and a desire was expressed to build a new sociological theory, whereby the bio-physical variables would play a definite and central role. In this regard, the contribution of Dunlap in 1970s and Catton in 1980s cannot be ignored.

Three seminal ideas were thrown up for consideration in the new paradigm along with the ecological critique of classical sociological theory. First, it was argued that modern sociology was greatly influenced by the social facts and by the notion of 'exceptionalism/exemptionalism' of the classical theorists. Secondly, mainstream classical and contemporary sociology, though apparently appearing as diverse, is basically nothing but a unitary paradigm (exemptionalist paradigm) whereas environmental sociology is a distinct and a competing or alternative paradigm. Thirdly, it was argued that one of the key tasks of environmental sociology is to provide a sounder sociological theory with emphasis that societies and bio-physical environments exist in reciprocal interaction with each other (Buttel, 2000).

Buttel (2000) further argues that environmental sociology will have a set of perspectives which cannot be divorced from classical theorists and have an ecological/material entology of some sort. For example, Marx and Engels emphasize that the importance of the sphere of production and social labour cannot be understood without their relation with nature. Even neo-Marxist Dickens (1992) emphasizes on the need for a political economy to treat relations between society and nature as complimentary to each other. Parsons (1977) and Burkett (1996) observed that Marx was fascinated by agronomy and the biology of agricultural production and in 'Capital', he had used agronomic examples to demonstrate the role of nature in social production. Like Marx, Durkheim also provided a modified evolutionary framework. He based his metaphors/assumptions on Darwinian evolution and organismic biology to provide a theory of social change, which gave direction from a low division of labour in primitive or simple societies to the complex division of labour in modern societies. This change of division of labour in society to explain the transition from mechanical to organic solidarity highlights population density, resources density and the competition for survival and these have a strong resemblance to a more modern notion of human ecology.

Weber was not in favour of a unilinear course of societal development like other evolutionary thinkers, but was of the opinion that social change was more a phenomenon determined by shifting the combination of subjective, structural and technological forces that were rooted in human motivation in history. His seminal work (*The Agrarian Sociology of Asian Civilization* and his general economic history) clearly brings out the impact of social structure on natural resources or the impact of natural resources on social organization.

The neo-Weberian environmental sociologists, West (1984) and Murphy (1994), have thrown light on environmental factors based on a sociological insight in their historical sociology of religion and ancient society. West (1984) considered that Weber in his analysis treated environmental factors as interacting in complex causal models, which favoured the selective survival of certain strata over others, an idea close to Darwinian thinking and highlighting the concrete instances of struggle over resources; for example, control of irrigation systems. Murphy (1994), on the other hand, based on Weber's concept of rationalization and his ideal types of rationality and orientations to action (based on his analysis of economy and society (1968)) argues that rationalization and the expansion of formal rationality may lead to an ethic of mastery over nature or to the quest for technology to realize this mastery (plasticity/flexibility of relationship between humans and environment) and to the lack of attention of human threats to the environment. He is of the opinion that ecological irrationalities may be caused by such rationalizations which may encourage social movements aimed at 'de-rationalization' or 're-rationalization' of modern institutions.

If we examine environmental sociology from the classical perspective in a historical sense, we confront two contradictory observations. On the one hand, environmental sociology appears to be materialistic as its thrust is on the natural world and the material sub-stratum of human life. Besides, it is objectivist as it focuses on the negative consequences of science and technology in explaining ecological and environmental sciences. On the other hand, environmental sociology has liberated and articulated ecology movements of various types. It is observed that the most significant and influential works in this area have tried to incorporate and synthesize both materialism and emancipation or liberation. However, some scholars have been critical of the classical tradition as these highlighted only negative social implications such as *anomie,* the iron cage and the disenchantment of the bourgeoisie society. Even Giddens (1987) pointed out that the classical tradition was largely pre-occupied with changes in the economy and in the structure of production as prime factors in shaping social change in the modern societies. Despite criticism of the classical tradition, environmental sociologists appear to be influenced by many of the concepts, methodological principles and pre-suppositions of this tradition. It is observed that lesser attention paid to environment in sociological inquiry may be attributed more so to the historical sequence of establishment of sociology as a discipline, to distinguish it from other natural and social sciences and as a sort of a reaction against Social Darwinism and geographical/environmental determinism. Even the enlightenment thinkers who influenced the Western culture for nearly 500 years, as well as its colonial and post-colonial expansionism did not pay enough attention to the environment. However, there are some definite limitations of classical theory for understanding the environment. In the context of environmental sociology, these have been identified as euro-centricities and innocence about globalization. These are the two main limitations of the classical tradition for understanding environmental issues in totality (Buttel, 2000).

Theoretical Perspectives on Environmental Sociology

It is observed that the growth and development of social sciences which got rid of reductionism of biological and physical analogies provided a variety of explanatory schemes and methods and helped in exploring a variety of empirical problems. This departure from normal science helped in framing and testing specific hypothesis to explain societal-environmental relations.

The development of environmental sociology is around three main arguments. One, the seriousness of environmental crisis by scholars during 1980s (Catton, 1980; Catton and Dunlap, 1978; Schnaiberg, 1980). This was realized by them primarily on the basis of secondary literature on ecology popularized by persons such as Paul Ehrlich and Barry Commoner. This trend is still very strong in the present context. Second, attempts to theorise the institutional dynamics of modern industrial society by emphasizing on market processes, capitalist relations, industrial relations, urbanization, political democracy, corporatist structures, social norms and cultural values and science and technological innovation, which led to environmental degradation and crisis. Third, a hope thrown up by these theories on degradation that these dynamics can and would be negated or over-ridden through public value change and/or environmental movement and mobilization since these are viewed as necessary to tackle environmental crisis and act as main mechanisms of environmental improvement (ibid.).

Buttel (2000) also highlights Schnaiberg's (1980) and Schnaiberg and Gould's (1993) environmental sociology which has talked about environmental destruction as a result of 'treadmill of production' engine by the competitive character of capitalism, social dislocations of private accumulation and private and public policies and practices leading to an exponential, capital-intensive environment-degrading economic growth. Schnaiberg also spoke of the environmental movement as the main countervailing social mechanism for societies to improve their environmental performance. However, the main drawback of these theorists highlighting environmental sociology is to theorize environmental degradation than to theorize environmental improvement.

May be, it is relatively easy to theorise about environmental degradation based on historical and contemporary data to demonstrate the various institutional arrangements that have affected environmental degradation processes. Further, these environmental sociologists seem to be over-emphasizing the coherence of environmental movements to have environmental improvement through them in an exaggerated manner.

These shortcomings, however, led to new environmental sociological perspectives to support the perspectives on the 'new human ecology'. In an endeavour to build up social constructionism or the cultural aspect of sociology, environmental sociology may appear controversial but it has brought out relativization of environmental science or global environmental change. In fact, this approach has thrown up three important points. First, the realist account of environmental science and knowledge appears to be one-sided and in consistence with any reasonable sociological perspective that should take into account both the structure and agency as well as the material and the symbolic. Second, interpretive discursive sociologies of environmental knowledge appear to claim environmental risks, environmental ideologies and environmental politics as a part of environmental sociology by highlighting how and in what ways the discourses are structured and re-negotiated. Third, it is critical for environmental sociology to recognize the pivotal role that the scientization of protests, social movements and politics is playing. In addition to the social constructionist perspective, some scholars have emphasized to bring in the tools of cultural sociology within the realm of environmental sociology through the application of cultural sociological concepts or tools, such as environmentalism, ecology movements, green parties and so on. This has mainly come from the scholars drawn in from mainstream sociology and it appears to be a positive development.

During the last two decades or so, there has been development of theories on environmental sociology in Northern Europe which have focused primarily on phenomena such as green parties, ecology movements, relationship between science and environmental movements in industrial ecology and so on. These theories seem to emphasize more on environmental improvement than to explain the environmental degradation. Further, these tend to highlight that there are ongoing complex changes in the social values and behaviour, mainly drawn from environmental degradation and crisis, simultaneously tending to downplay the role of certain forms of environmentalism, especially radical, expressive environmentalism in environmental improvement.

Buttel (2000) opines that two emerging traditions in environmental sociology such as risk society and reflexive modernization have played a significant role in taking note of the shortcomings of environmental sociology by the American scholars. Both these theories are built around Beck et al. (1994) ideas of risk society and Anthony Giddens' concept of reflexive modernization highlighted by Goldblatt (1996).

Beck et al. (1994) have talked of reflexive modernization as a sort of transformation from an industrial society to a risk society. In his opinion, modernization has led to a set of risks and hazards that not only pose threats to the current generation but may also affect the quality of life and, more possibly, the very survival of future generations. The other major fallout of reflexive modernization is the decline of parliamentary institutions, the individuation of politics, the decline of class and the growing role of sub-politics. The risks and hazards generated by modernization or by the industrial society call the central institutions of society such as science and the legal system into question. Further, there is a need to free actors relative to social classes and social structural categories to build up new identities and roles and form new social movements around them about risks. In other words, citizens need to scrutinize science, technology and scientific institutions by making moral claims that are equal to those of modern science. In that sense, reflexive modernization does not mean less but more modernity radicalized against the paths and categories of the classical industrial tradition (Beck, 1992).

The second theory which has emanated from reflexive modernization is teleological modernization theory. In this regard, some scholars seem to be critical of the way Beck has used reflexive modernization to highlight the risk society theory. But the two main thrust areas in this theory are: one, human and institutional choices are not simply the reflection of master structural forces of capitalism and industrialization and the second that the solution to environmental problems will be in the progressive modernization of societies, rather than in the de-modernization or counter-modernization suggested by radical environmentalism. The seminal idea of this theory projects an optimistic view of the potential for technological change, which may lead to solutions for environmental problems. In fact, it is expressed that at least in some societies, the socio-economic influences on technological research and development and industrial technology choice may result in improvement in the efficiency of conversion of raw materials into finished products so as to affect reduction in the quantity and toxicity of the waste stream from industry. Further, this theory holds good for the State Environmental Policy; for example, new trends in the structure of states which advocate decline of command and control, environmental regulation, etc., resulting in modernization of the state are, in fact, stimulating environmental self-regulation within the civil society through legal mechanisms of civil liability law, economic mechanism of market-induced efficiency and environmental policy incentive and citizen movement pressures (Buttel, 2000).

As risk society and ecological modernization theories gained momentum, they invited several kind of criticisms, especially from the perspective of developing countries. These theories are mainly Euro-centric and lack generalizations which may apply to most societies in the world. More specifically, in the context of risk society theory, it is argued that the focus is on its quasi-constructivist conceptualization of technological and ecological risks. This appears to be less noticed and pervasive in the case of most societies as a concern of public pre-occupation. The other major limitation of this theory is in regard to the notion of the equality of risk it portrays, as without class dimension, one cannot exempt a person from large-scale hazards and risks. It is observed that in most societies of the world today, there is relatively little equality of risk and environmental equality is the rule rather than the exception. In this regard, Murphy (1994), while recognizing the highly unequal distribution of costs and benefits of environmental extraction and destruction among various social groups in the United States, has thrown up the notion of environmental classes. This is based on the inequality of risks, hazards, access to resources, consumption capacity and other environmental processes as sub-texts of environmental sociology.

The ecological modernization theory appears to be more influential than risk society theory because it is positive for the environmental sociology to be socially useful and suggests possibilities for environmental improvement. Further, it provides right portrayal of the cultural processes and political structures of state-induced environmental reforms in most parts of the West. However, despite its lasting influence on environmental sociology, it has also been subjected to three main criticisms. First, the premise that social and political institutions can provide strong signals to industries and consumers to be pro-environmental friendly and in the absence of that, it would lead to costly public or private sanctions. But, this premise is not only a 'law book' for the Western world but is also being pursued and followed by developing countries, which have also framed environmental laws and regulations, including customary social relations and practices such as common property regimes, to conserve resources. Secondly, it is observed that the types of environmental reforms which appear to be progressive such as pollution control and source reduction, etc., do not result in decrease in the human toll on environmental resources, since despite such efforts, aggregate consumption of minerals and other raw materials over the past two decades has remained unabated. Thirdly, it is also noted that this perspective may legitimize a political culture of environmental policy-making that may absolve industrial corporations and other agents of environmental destruction of their responsibilities, while ignoring the fact that the modernization of the state is contributing to growing inequalities of wealth and power within countries and across the world system. In order to improve the ecological modernization theory on environmental sociology, perhaps, there is a need to address some of these criticisms and include concepts such as embedded economy and state-society synergy as advocated by Peter Evans (1995; 1996). The state-society synergy has been employed in the context of environmental sustainability issues, especially in the context of urban sustainability and livable cities in the nations of the South (Evans, 1998). This work provides a model of states and power relations which is not devoid of coercion and domination, but can also have positive impacts on and through the civil society process. Further, his observations on the role of state-society synergy at the sub-national level is worth noting; the way the political cultures and configurations can have positive impacts on environmental improvement in the developing nations (Buttel, 2000).

Goldblatt (1996), in his analysis of social theory and environment, has not shown much enthusiasm about the contribution of classical thinkers like Marx, Durkheim and Weber in reference to relationships between human societies and natural environments. Though Marx tried to relate human life with the natural environment, his analysis of capitalist industrialization shifted agricultural economies to the periphery or to a stage of marginalization. Durkheim, on the other hand, took into account the population density and its relationship with material resources in the context of evolutionary stratification of human societies, thus making natural world a significant causal factor in human history. Weber took away sociology from the influence and determinism of biology and his engagement with the natural world has also been not that significant. As such, it is observed that economic capitalism was not seen to be environmentally damaging in any big way for the people on this planet and especially those living in the developing nations.

However, Goldblatt (1996) is of the opinion that the contemporary social theorists such as Giddens, Gorz, Habermas and Beck have through their critical investigations highlighted that environmental degradations are generated by the interaction of economic demographic, cultural and political factors. Out of these, economic power is most significant and can be best analysed through the model of capitalist rather than industrial economics. As regards the rise of environmental politics, it should be based on historically specific, actual and perceived transformation of the environment to raise environmental movements as a result of the material progress. These transformations should re-define material interest and moral ideas because beyond a certain point of material prosperity, they can remain only as unorganized and unarticulated protests. Also, the structural impediments to the political success of such a movement may turn out to be its greatest opponent just as material prosperity may give rise to environmental politics, which may also act as its opposition. The changes in the consumption pattern and definition of well-being for bringing the Western societies within the realm of sustainability may be extensive in their coverage and intensive in their consequences. However, in this transition process, everybody is likely to be affected and that may result in making negotiated social change as a big political task. But, the political and legal systems of capitalist societies are not neutral but are biased towards allocation of power to environmental interests (ibid.).

Goldblatt's argument is that though Marx's conception of socialism brings in the moral dimension of environmental degradation in the cultural context, Habermas is critical of the weakness of such a concern because of its productivist paradigm, holistic conception of society that underestimates the factious nature of relationship between social stratification and political agency, inadequate theory of modern state, underestimation of the importance of democracy in complex societies and its evolutionary model of history, having unacceptable normative presuppositions supporting unprincipled vanguardism. In the wake of that, Habermas strongly advocates a participatory debate in a more radical democratic form at the widest level than what is existing in the Western societies; in order to have moral articulation and power, the requirement to place the implications, causes and consequences of environmental degradation on the political agenda in a realistic manner.

On the other hand, Giddens, while discussing his project of late modernity, is hesitant to accept a simple relationship between cognitive capacity and political potential, as he considers that more knowledge does not necessarily mean more control because it is more knowledge in science that has led to environmental degradation. He also thinks that due to collapse of the socialist world, a political paradox has emerged on the scene in the sense that radical socialism has become conservative while conservatism has become radical. As such, the socialist perspective to direct social change because of contemporary economic limitations and political fortunes has lost its significance, mainly because now modern societies are getting transformed by the intersection of globalization, de-traditionalisation and manufactured uncertainty or risk, which form the basis of reflexive modernization. In addition to radical politics based on the elements of socialism and conservatism, he advocates dialogical democracy in response to environmental degradation problems and other kinds of social changes, including moral dimensions, on the basis of consensual agreement rather than through the use of threat of extraneous force. Beck, though, has not dealt that strongly with the decline of socialism but his arguments do reinforce Giddens' position. Gorz is also of the opinion that we should not rely heavily on classical socialism and look for a political and economic system having many social situations and processes of self-definitions to build up self-identity, rather than depending on the conception of well-being of old socialism. He is for a political economy that should provide liberation of time in terms of massive reduction in working hours, increased flexibility and control over time, expansion of unhindered free time and unfettered logic of economic rationalism. These can be best realized by radicalized trade unions rather than by the old socialist model of working class mobilization and political action in environmental movements.

In the wake of the above commentary on these modern theorists, Goldblatt (1996) raises certain questions such as whether or not we can vouch for protection from the threatened life world through the processes for social and cultural rationalization as a result of transfer of economic and political action from the system depending on it, or can we create a fully radical and dialogical democracy to garner greater participation in the decision-making. And finally, whether or not we can do away with representative democracy and state power for instituting social and economic changes. In the case of complex societies which are launched, participation in public spheres or debates or ecological democracy is always problematic due to simultaneous operations of transnational, cosmopolitan and global democratic institutions.

Environment, Ecological Fragility and Sustainable Development

The Environmental Debate

The environmental debate mainly centred around the environmental crisis created by overexploitation of natural resources through the processes of industrialization and consequent urbanization. One school of thought vouches that a crisis does exist in the sense that there is a threat to man's survival. In fact, due to environmental degradation, decent living of mankind is becoming difficult. The damage caused due to overexploitation of resources and extreme over-industrialization is so severe that unless some drastic remedies are taken, there is hardly any hope to sustain life on a continued basis. The damage to the environment is already affecting the life support system on a global scale and may result in the breakdown of society and lead to famine and disease within the lifetime of children as of today. On the other hand, there are those who contend that this is a pessimistic approach and we need to rectify the errors involved in our technological development. We do have reserves of materials and energy and we can depend upon our resilience in the life support system. This may show positive signals for optimism about the future.

Notwithstanding this debate, which became alive during the mid-1960s by some prophetic voices who caught the attention of public at large, this warning about the dangers of the future was pronounced at a time when the natural environment was showing not so severe signs of damage. However, we cannot abide by the fact that environmental damage is only a phenomenon of the last 200 years or the primitive man was well adjusted to his natural environment and he might have damaged it while making mistakes in his adoption or adaptation to the environment, because he lacked the skill or knowledge. In fact, there might have been severe damage to the fertility of the soil and destruction of forests. Further, in the early man era, there might have been extermination of complete species of animals in a much slower manner than what we observe today over long periods of time. In fact, before the industrial revolution, there were not many debatable points raised in the context of environmental crisis, but we had some debate about the nature and behaviour of man and his relation to the natural world and to other men. However, in the context of the modern era, we have enormous scientific and technical developments. We could gather data, communicate and manipulate these with speed and facility and conduct studies. This was perhaps not possible fifty years ago. One voice that predated the early years of this period was that of Malthus.

It was towards the late 18th century when the population of Western Europe was passing through its growth period that a young British scholar, Thomas Robert Malthus, published two essays on the Principles of Population (1798). He was often misquoted and misunderstood by his contemporaries because he had made a forecast of imminent catastrophe which may come directly from the pressure of population. Malthus (a Cambridge mathematics graduate) built up his argument on the premise that man needs food to survive and that human sexual passion is likely to be a constant feature throughout man's history. He argued that populations will tend to grow unless they are checked by effective ways in areas where food was not available. Moreover, the power of population growth was likely to be greater than the capacity for food production. His one-line theory is while food production grows arithmetically, population grows geometrically. In the event of the population rising, it might require some checks based on deliberate choice in the form of moral restraint, misery and vice through war, famine, disease, hardship, extra-marital intercourse and the use of contraceptives. He was more in favour of preventive checks like moral restraint than checks such as misery, etc. But, Malthus was proved wrong because the scientific developments in the agricultural field kept the per capita food production at a more satisfactory level than he would have thought. He also failed to foresee the medical revolution that brought down the death rates in a dramatic manner and consequently increased the power of population. Many people in the present scenario may not agree with these assumptions for his definition of vice and his total rejection of contraceptive devices, but we are fully knowledgeable now about the capacity of the Earth to produce subsistence levels of food and minerals. The capacity remains finite and is likely to be threatened in the long term by the short-term techniques which may cause disturbance in the ecosystem. However, he was optimistic about man practising self-restraint and focused on the responsible role of governments in controlling populations. However, his ideas were treated with pessimism and in abusive terms (Harvey and Hallett, 1977).

After the publication of the second essay on population by Malthus in 1803, though in the mid-1960s there were many contributions to the debate on environment, yet the meaningful publication that came as a book was entitled *Man and Nature, Physical Geography as Modified by Man* by a geographer George Perkins March in 1864. In fact, this book was revised several years later as 'Earth Modified by Human Action'. This geographer became influential in starting the campaign in America to conserve natural life and beauty. Conservation may not be considered the same as global ecology, but it showed that we are destroying something on which we depend for survival After the Second World War, people got concerned about the future of man and many American writers saw the dangers of a situation where growth and development were not matched with the understanding and care for the land. The outstanding contributions in this regard were William Vogt's 'The Road to Survival' (1948), while Fairfield Osborn's 'Our Plundered Planet' (1948) appeared alarmist on the agricultural practice of that time. Later, these scholars also contributed significantly through their works to the debate to draw public attention and concern about the environment. Vogt wrote 'People's Challenge to Survival' and Osborn 'The Limits to the Earth' (as quoted by Harvey and Hallett, 1977).

Ecological Fragility

Later on, Berry Commoner, an American biologist, in his famous book *The Closing Circle* (1963), contends that the root cause of environmental degradation is that the technologies employed during the last 75 years have been ecologically faulty and their use for long has resulted in creating stress on our life-support system and the destruction of our biological capital. The damage caused is so high that it poses a threat to continued development. Our priorities must be to repair the damage already done and replace present technologies with those that are ecologically wiser. He provided a quantitative relationship between population, affluence and technology and observed that the new technology is an economic success because it is only an ecological failure. Taking the example of fertilizer, nitrogen-linked with food production in USA in 1949, he arrived at the conclusion that increasing the technology factor is far more important than the small rise in population so far as its contribution to environmental damage is concerned. Further, the environmental impact is very large than the benefit in almost all the examples he studied.

The other contribution by Paul Ehrlich (1972), though critical of Commoner, is worth noting. He considers Commoner's work as one-dimensional ecology since he took the base data in 1900 and treated it 75 years following that, for the purpose of making data more accessible and reliable. However, his methodology has hidden the truth that serious ecological impact, in fact, began 10,000 years before when man moved into settled agriculture. He is of the opinion that the sheer pressure of numbers, the inertia of population growth and the social consequences of the threat of famine and disease are going to be significant in causing impending eco-catastrophy. He, along with his wife, dealt with these aspects in a major way, in their work *Population, Resources, Environment* (ibid.). No doubt, he is also indicating misguided technology and makes frequent references to other growth than population alone. He is a strong advocate of population control as he feels that this can lead to halting of other human problems such as poverty, racial tension, urban blight, environmental decay and warfare, etc. Though he does not consider population as a panacea to solve all these problems, yet he opines that whatever may be the cause, it is a lost cause without population control. In fact, the husband-wife duo highlight that social and technical activity is more vital for the survival of mankind than what Commoner had worked for a change in the USA as well as for fighting famine and malnutrition in the Third World countries. They also express the need to support those politicians who show some kind of awareness and concern and take a kind of stand on the environmental issues.

As a reaction to these popular presentations, a number of critical articles appeared focusing on examples of errors of data and the relationships established in them. A book by Rachel Carson (*Silent Spring,* 1963), a specialist scientist, needs special mention. She highlighted that DDT and related pesticides not only kill insects, pests but many other organisms as well and endanger the life support system as a whole. This has been confirmed by subsequent research in this field despite the errors observed in her research work. Though the environmentalists do acknowledge this kind of criticism, yet it is argued in certain quarters that this analysis may present only the best possible picture for decisions on environmental policy to be made, to have a perfect case for a change. Further, it is argued that it is not necessary to make any policy change as proposals for reform need greater scrutiny than reasons for accepting the status quo.

One of the arguments is that there is hardly any problem that cannot be dealt with by a more skillful use of science and technology. Besides, economic growth is necessary to provide resources and goods to meet the basic needs and people can attend to such matters as environmental degradation and the quality of life. Another argument is that any vigorous campaigner for environmental issues has ulterior motives. In other words, there is a suspicion raised about him or her for having a desire to show that his or her class of nation should suffer least in the future if things become difficult. This type of observation that has come mainly from the Third World commentators is of no surprise due to many years of frustration caused by the growing gap between the rich and poor. The environmentalists in these nations perceive that there was not much available as was thought earlier. As such, it may not be wise to go for industrial development and it may be better to have a growing population under control to reduce the pressure on the available resources (Harvey and Hallett, 1977).

There is no doubt that the human ecological fragility is a result of extreme economic growth of industrialization. In fact, it was so enormous that it crossed the limits of growth and involved unseen distortions, misuse and waste. In this context, there are two significant reports – 'The Limits to Growth' and a 'Blueprint for Survival' – which need to be discussed briefly.

'The Limits to Growth' was brought out in February 1972, based on the report of the Club of Rome project on the 'Predicament of Mankind'. This report was an outcome of cooperation between a group of international consultants under the chairmanship of Aurelio Peccei and an international team of scientists and computer experts at the Massachusetts Institute of Technology (MIT) who had come out with a new approach in the computer modelling of complex problems. The Club of Rome had its first meeting in 1968 with 30 individuals from ten countries and over a period of two years, their number grew to about 75. Several meetings resulted in the identification of 'world problemitique', indicating a complex web of technical, economic, social and political problems faced by all countries in the world. It was Professor J. Forrester of MIT who had developed a technique of mathematical modelling through the use of computer for the analysis of these interacting problems identified above. He was later joined by Dennis Meadows and his team. (The funding of this project was done by the Ford Foundation and later by the Volkswagen Foundation). They chose five basic quantities or variables which may indicate the state of the world system, in the best possible way. These are population, pollution, natural resources, agricultural capital or output and industrial capital or output.

After using 1970 values as a basis, they tried to establish levels and rates of flow along with feedback mechanisms in order to describe their best interrelations. The standard world model brought out that food, industrial output and population grow exponentially until the rapidly diminishing resource base forces a slowdown in industrial growth. Due to natural delays in the system, both population and pollution continue to increase for sometime after the peak of industrialization. Population growth is finally halted by a rise in death rate owing to decreased food and medical services. It was also brought out that the world was finite and that growth of these five quantities could not continue indefinitely. They were looking for a smooth transition from a growth to an equilibrium condition in which population and the quality of the environment attain stable and sustainable levels.

This report had support as well as criticism. The developed countries' economists and politicians criticized the desirability of non-growth or a steady state economy. In fact, the whole notion of limits to growth was under attack. They challenged the basic idea of the existence of physical limits and showed confidence in the possibility of man to carry on 'unlimited technological migration', that is, moving on from one technology to the next as the need arises. The Third World nations provided a different kind of criticism, not that much on the technicalities but on the motives of the Club of Rome to make easier the exploitation of resources by multinational companies and to carry forward the existing unjust distribution of wealth and power. In response to these criticisms, the Club of Rome came out with a second report in 1975 (entitled 'Mankind at the Turning Point'), based on the computer world model employed earlier to carry forward the analysis while dealing with the objections raised by critics.

The Ecologist magazine (January 1972) devoted the entire issue as a blueprint for survival by eminent scholars and they tried to go beyond the analysis of physical and economic aspects and brought out a new philosophy that could be sustained for the indefinite future. In fact, they put forward a comprehensive programme for bringing about this social order

A 'Blueprint for Survival' which later got published as a book brought out four main conditions of a stable society. One, minimum disruption of ecological processes; two, maximum conservation of materials and energy or an economy of stock rather than flow; three, population in which recruitment equals loss, and four, a resource system in which the individual can enjoy rather than feel restricted by the first three conditions. It was observed that these could be achieved through seven operations:

1. A control mechanism by which environmental disruption is minimized by technical means.
2. By having a freeze operation in which present trends are halted.
3. Substituting the most dangerous components of these trends by technological substitutes or innovations having less deleterious effects.
4. Substituting these technical substitutes by natural or self-regulating ones, which may not cause undesirable disturbances in the normal processes of eco-sphere and can become sustainable over very long periods of time.
5. The invention, promotion and application of alternative technologies which are energy and material conservative

(for example, intermediate technologies for small economies, which disrupt ecological processes in a minimum fashion).

6. Decentralization of polity and economy at all levels by having formation of communities, small in size so that they could become reasonably self-regulating and self-supporting.
7. Education for such communities.

These developments or environmental debates lent support to the United Nations Conference on the Human Environment in Stockholm during the summer of 1972 and the United Nations designated it as a World Environment Year. The UN commissioned Rene Dubos (French human ecologist who gave the slogan of 'Think Globally but Act Locally') to collect relevant material on the subject, which was distributed to all the participants and who later on brought out a book jointly with Barbara Ward (*Only One Earth,* 1972). This Stockholm conference was followed by UN conferences on the use of the sea, on food and agriculture and on population. However, in these conferences, one finds an enormous gap between the powerful speeches of the national delegates and the inactivity of governments for taking unilateral action where agreement cannot be reached on concerted action by all the involved nations (as quoted by Harvey and Hallett, 1977).

As brought out in the above debates and counter-debates and discussions among scientists and environmentalists in conferences, etc., on the environmental issues, one thing that appears to be significantly ignored is that the resources on planet Earth are finite and these need to be exploited for the development of societies in a cautious manner so as to provide minimum decent living to the inhabiting populations in different nation-states, not only for people in the present time but also for that of the future generations.

In view of the above, the world leaders felt that due to over-industrialization of nations because of burning of fossil fuels to produce goods and services and as a result of intensive and extensive use of chemical fertilizers, pesticides and fungicides to raise agricultural productivity per hectare of cultivation for different crops, the environment has been affected adversely due to the pollution of air, water, soil erosion, desertification, waterlogging, salinity, ground water depletion, etc. As such, they made a clarion call for sustainable development of the developed as well as developing societies.

Need for Sustainable Development

Maxneef (1990) identified three main semiological phases in human development. The world economic crisis in the late 1920s and early 1930s was well-tackled by employing Keynesian macro economics, through the input-output growth strategy. However, it was Schumpeter who conceived the concept of economic development as a development language during the 1950s, especially in relation to the reconstruction of post-war Europe. The underlying assumption in his paradigm was that economic growth will eradicate poverty across the world through industrialization, urbanization and modernization, big push forward and 'take-off' into self sustained growth. This approach continued till the mid-1970s but from mid-1970s and 1980s, it was felt that the developmental inputs for the generation of growth are not producing desirable effects. A mega crisis was being faced by the globe with regard to unlimited economic growth and expression on one side and that of the social-ecological collapse on the other. As such, efforts were directed to bridge economics and ecology with social life and these efforts were being referred to as sustainable development at the global level. In this process, however, the developing nations which are getting technological benefits appear to be the worst-affected as they, in comparison to the developed world, do not have sufficient know-how of clean technology so as to take care of their environmental hazards. Further, there are limitations on their part to make use of technological innovations evolved by the developed world to their advantage.

This brings us to the conception of sustainable development in a more elaborate manner. Though the origin of the term is traced to the World Conservation Strategy Document, the concept became popular after the report of the World Commission on Environment and Development (Brundtland

Report). This document defines sustainable development as the path of human progress that meets the needs and aspirations of the present generation without compromising the ability of future generations to meet their needs. It is not that easy to accomplish this because we need a basic change in our mindsets, that is, the way we think, live, produce, consume and the way in which we behave towards each other.

It is a complex phenomenon and we need to pursue multi-level processes in a systematic manner to achieve it. Lender (1989) suggests a system approach for that. It should focus on:

1. A political system ensuring effective citizen participation in decision-making on environmental issues.
2. A social system that should provide solutions for the tensions generated by the disharmonious development process.
3. An economic system that produces surpluses through technical knowledge on a self-reliant sustained basis.
4. Sound production system to preserve the ecological base for development.
5. Technological system that should provide continuous search for new solutions to environmental problems.
6. Administrative systems that are flexible and self-corrective to take stock of environmental degradation, and
7. An international system that should foster sustainable patterns of trade and finance.

Further, it is observed that the sustainable development efforts have to be dynamic and not static. These should be ever changing and constantly requiring adaptation to various systems for meeting human needs and aspirations, not for one country or region but for the global population.

The challenges of sustainable development were brought out in the United Nations Conference on Environment and Development (1972) at Rio de Janeiro (Earth Summit). It was emphasized that without adequate and better environmental measures, the development process will be undermined and with accelerated development in poor countries, environmental policies will fail. We need to have a trans-disciplinary approach in tackling the issues involved in sustainable development. Economists must relearn the significance of natural capital along with the maximization of human welfare within the limited existing capital stock and technologies. Ecologists will have to preserve all ecosystems to the desired level and subject these to the natural life support system, while still maintaining the integrity of various sub-ecological systems. In these processes, the key factor is the human actor. It is highlighted by sociologists that the social components of sustainability are as significant as that of economic and technical aspects because the patterns of social organization of human beings are crucial for finding viable solutions in achieving sustainable development (Serageldin, 1994; Cernea, 1994).

Ecological and Economic Perspectives on Sustainability

With regard to the definition of sustainability, it has been argued that it must embody the idea of a forward-looking community that is normatively respectful of the pursuit and also engaged in the perpetuation of knowledge as a path of truth-seeking (Peirce, 1960, as quoted by Norton, 2003). As the focus of the study is on environmental sustainability, it may be apt to briefly highlight the debate on sustainability that has come from ecologists and economists. There are theoretical disagreements that cut across disciplinary boundaries and because of that, the term 'sustainable' becomes conceptually ambiguous as both the economists and the ecologists emphasize on different aspects in explaining the interaction of humans with their environment. In view of this, it is rather necessary that scientists across the disciplines have better communication and convey the agreed dialogue of information to the decision-makers for the right kind of interpretation of data.

Though a new paradigm of ecological economics is emerging, economics as a part of mainstream discipline is basically involved in understanding the values of the public and their behavioural consequences. The focus is to formulate questions so as to describe values held and to express concepts that can influence real decisions. In order to describe and record social values or to evolve an accounting system for sustainability, we have to base them upon perspectives that may appear controversial in the interdisciplinary discussions on environmental policy.

We do have two clusters of issues that intersect at the edges of economic and ecological science, having a bearing upon environmental policy and its practice. These are the problems of 'reversibility and substitutability' and the accounting problem. The main disagreements about the concepts of reversibility and substitutability among resource ecological systems and their components are due to different attitudes of economists and ecologists towards discontinuities and thresholds in understanding and managing the impact of human culture on an ecological system.

For economists, substitutability means the capacity to alter consumption and production activities in the event of increasing scarcity of some resource in order to maintain a desired overall flow of services, while reversibility means economic consequences broadly defined to include both market and non-market values rather than to the physical states of ecosystems. However, ecologists refer these terms more so to physical properties of the ecosystems. For them, reversibility is related to the resilience of the ecosystem, that is, its capacity to return to high level of fractions after being disturbed. They consider substitutability as a kind of redundancy in the sense that if some attribute of a system is diminished, then one has to look for other sources of that attribute. In fact, for them, substitutability becomes important if natural degradation is large in scale. However, economists are seized with sustainability in terms of maintaining high expected levels of human well-being over time. As such, they show concern over the capacity of natural environment and other social assets to meet human wants and needs. This includes a variety of preservation and bequest, besides motives to raise interest in the use of the ecosystem or that of resource consumption. The data on depletable resources and economic progress bring out that there is relatively a substantial capacity to substitute for diminished capital services as that would be required to maintain consumption of final goods and services over time. This appears to be inconsistent with physical laws. In contrast to the economic perspective of ecosystems as service factories, ecologists view them as complex, dynamic sets of resources that have been organized on multiple scale bases (Norton, 1955).

It is argued by Holling (1996) that the substitution paradigm in economics is suitable for lower scale ecological systems as providers of services. But, in large scale systems such as forest systems, the response will be very slow and with less redundancy. Even the resiliency of the system will move up to a point beyond which it cannot be pushed and that may result in rapid discontinuous changes in the function of the system and may require much more time for recovery. In the above backdrop, ecological theory and economic analysis may require a treatment of interactions, amongst scales in a system.

It is also argued in certain quarters, for example Solow (1993), who comes to the conclusion that though we owe the future opportunities equal to our own for the next generations, but to do so we shall have to maintain a non-declining stock of aggregate capital. Only then we can discharge our obligation to the future by maintaining adequate investment so as to compensate the future for the use of degradation of particular resources. This is the kind of unstructured bequest package between generations and we can treat sustainability as a sort of obligation to honour a fair savings rate across generations. But, this argument is not acceptable to those who are in favour of preservation of a variety of ecosystem functions to enhance the well-being of future generations. They are advocating an ethical dimension to protect these processes so as to provide adequate options to future generations. They are in favour of a more highly structured bequest package that should take into account the natural and man-made capital separately with special attention to the former. The basic point of disagreement between ecologists and economists is the problem of environmental accounting in terms of assigning values, types of values, measurement dilemmas with regard to changes in the states of the world that may result from human action to degrade or protect the environment. If sustainability is agreed to be conceived as a concern for the future, including horizons beyond the length of a human generation, then we also have to come to some common understanding across different disciplines of humanities and social sciences regarding the formulation and evaluation of human values for the well-being of future generation or our obligations towards them (Norton, 1955).

Unfortunately, no discipline such as ecology, economics or philosophy has been able to provide a coherent and complete understanding of human values across multiple generations. Though economists do not suggest a single approach in assessing inter-temporal distributions of well-being, yet they argue that some kind of discounting is necessary for the time preferences exhibited in individual behaviour and also to conceive potential for the future to be better than the present as a consequence of economic growth. However, many of them do not support this standard approach to discounting for inter-generational welfare trade-offs as distinct from inter-temporal allocations across one's life. In fact, there is a need to define a social discount rate for inter-generational allocations which should be different from the rate of personal time preference. However, Burton (1993) suggests that the social rate of discount is a sort of reflection of societal value judgments with regard to appropriate inter-generational distribution of well-being, including the potential for future generations to be better-off than the present from the economic progress or worse-off due to environmental harm. This appears to be a better concept from the rate of individual time preference because by using it, both positive and negative effects can be evaluated in terms of their consequences for utility purposes (Norton and Toman, 1997).

For sustainable development, we have to emphasize on an improvement in both the quality of life and material security of the more than one million poorest on this planet, along with the reversal of present environmental degradations. In order to accomplish that, we have to promote economic and environmental interdependence across the globe to build co-operation between men and women in this interdependent world and make a shift from an ego-centric human life to God-centric vision of life (Salim, 1991).

In the larger interest of humanity on this planet, in order to provide better life chances for our future generations, we have to give importance to what Mahatma Gandhi preached. 'There is sufficiency in this world for man's need but there is not enough for man's greed'. In fact, need rather than greed of man should be the guiding principle for sustainable environmental planning and policy.

Relationship between Environment, Health and Population

In the preceding pages, an attempt has been made to throw up conceptual and methodological issues involved in analysing environment and ecology (natural vis-à-vis social), environment and technology, environment and economic development, environment and its relationship with sociology in the context of classical theory and contemporary environmental sociology, ecological fragility caused by environmental overexploitation, need for sustainable development with emphasis on economic and ecological perspectives in sustainability.

The focus of the present study will be to highlight the relationship between environment, health and population. Environment, whether physical (natural) or social and cultural, if not protected adequately may cause health hazards to the inhabiting populations. In a presentation made earlier by analysing the relationship between environment and economic development, it has been clearly brought out that resources on planet earth are finite and efforts to overexploit them through technological breakthroughs and innovations may lead to many kinds of hazardous results. There is always a limit to growth and if we have to provide minimum decent living conditions to the present as well as future generations, we shall have to act haltingly and cautiously to conserve and preserve our resources, otherwise the consequences for the populations in different countries of the globe will have damaging effects. We are already experiencing a wider and intense phenomena of air pollution, water pollution, soil degradation, depletion of ozone layer causing health risks of various kinds, species extinction, deforestation causing floods and landslides, waterlogging, ground water depletion, wasteland formation, etc., due to extensive fossil fuel burning and intensive use of chemical fertilizers and pesticides, cutting of forests, etc., and effects of green house gases having adverse effects on human life.

The environmental damage caused has serious problems for environmental sustainability, especially for the developing nations as unlike the developed nations, they do not have either clean technology to tackle environmental crisis, or nuclear energy generation resources to take care of their infrastructural, industrial, commercial and agricultural activities, including animal husbandry and fishing, which are ever increasing and expanding in response to the demands of a rise in population. In this process of prosperity and economic pursuits, not only the environmental surroundings but the health of the people assumes great significance for production of various goods and services as well as for raising the productivity of man per unit. We need to go beyond the game of numbers in having quality of population. This brings us close to the health aspect of the population.

In order to understand the relationship between environment, health and population, an attempt may be made here to analyse them within the perspectives of environmental sociology as highlighted in the preceding paragraphs. In analysing environment within the sociological perspectives, it is clear that we cannot dismiss classical thinkers like Marx, Weber and Durkheim. It may not be possible for us to think about the various spheres of production and social labour involved as divorced from nature. This is what Marx has emphasized in his analysis of modes of production and relations based on that. Even Durkheim's contribution, especially the concept of division of labour, has an influence on human ecology while explaining the shift from mechanical to organic solidarity. This shift has highlighted population density, resource scarcity and competition for survival having a bearing on modern notions of human ecology. Weber also provided material in his general economic history on the impact of social structure on natural resources and vice versa. Even his concept of rational orientation and watertight iron cage bureaucratic structure has some implications for analysing environmental policy and activism, because many of the political decisions by the governments do not translate into action to conserve, preserve and save the environment.

Modernization paradigm by sociologists for growth and development was based on heavy inputs of industrialization and consequent urbanization, not only for the developed but also for the developing nations. This paradigm did not take into account the environmental hazards and destruction caused by the abundant use of inanimate energy generated through intensive burning of fossil fuels like oil, coal and natural gas and even nuclear energy. In addition, agriculture production was raised by the intensive use of chemical fertilizers and pesticides, which resulted in causing health hazards to the people engaged in agricultural operations. These processes of growth and development led to impoverishment of a large proportion of people, especially industrial workers and agricultural labourers. They had to bear the brunt of industrial emissions, chemical waste and effluents and spillover of oil into the river and sea water. Further, the toxicity of agricultural produce and plants has affected the health of the populations, especially those engaged in the production processes, without being provided with any preventive and protective measures. They become 'at risk' and vulnerable to all kinds of ailments and diseases, especially air borne and water borne, due to rise in the quantity of viruses and bacteria in air and water.

The modernization paradigm and even the dependence theorists who advocated import substitution industrial development for the developing nations were responsible for environmental degradation in the absence of any improvement mechanism built in the environmental policy for promoting sustainable development. These paradigms created an elite class structure and marginalized a large proportion of people in the developing nations, who became victims of poverty, disease, hunger, overcrowding, insanitary living conditions, etc. No doubt, in response to such indecent living conditions, many environmental movements came to the forefront and showed concern through effective political voices, but not much could be achieved in this direction mainly because the implementation of decisions arrived at between the governmental and non-governmental organizations (NGOs) could not be carried forward.

Ecological modernization as a part of environmental sociology in the Reflexive Modernization Schema, especially by Beck, highlights the emergence of risk society that may get evolved as a consequence of hi-tech industrialization and capitalist economy. A set of risks and hazards that may not only threaten the quality of life of the present generation but may also risk the survival of future generations. For Beck, the ecological crisis caused by risks of mega accidents such as the Bhopal gas tragedy and Exxon Valdez, the environmental risks of recombinant DNA and genetic engineering (cloning), personal risks caused by medical high technology and failure or limited capacity of nation-states to directly regulate the production processes may have serious consequences for the human well-being.

Of the contemporary social theorists, there is a need to take into account the contributions made by Anthony Giddens and Habermas, as well as that of Gorz (Goldblatt, 1996). They are for dialogical democracy, fully radical in order to have greater participation of people in decision-making on the environmental degradation problems and other kind of social changes, including moral aspects, by a consensual agreement rather than by the use or threat of force. Some of these theoretical inputs outlined above will be suitable for analysing environmental issues and crises for mobilization of people at the helm of affairs in the management of environment, as well as those affected by them.

It is well recognized that population is a very crucial factor in development. There is close interaction between development and population growth as one may act as an antecedent to as well as a consequence for the other. In fact, reduction in population size may not always be desirable for every kind of development because in certain situations, there may be a need for huge manpower resources for development purposes. In the context of developing nations like ours, the rise in GNP has not resulted in reducing a large proportion of population affected by the poverty syndrome due to technological backwardness of agriculture and incapacity of even technologically-advanced industrial nations to absorb the fast-growing labour force as a result of population growth. This has resulted in an increased unemployment rate. Besides, the education and health care of developing nations could not be expanded because of rise in population as a necessary input for human resource development. Population growth has also forced these nations to go in for foreign aid and investment, thus promoting economic dependency on the developed nations.

We cannot set aside the geometrical and arithmetical postulates given by Thomas Malthus in the late 19th century, suggesting that food supplies will fall short of population rise. The technologists assure us that with appropriate technology and sufficient purchase inputs, it would be possible to provide 50–60 million people with high-quality diet or almost 100 million people could be fed at a minimum level of subsistence. This was brought out by Meadows and his associates in the Club of Rome report (1972). They had warned that the world would run out of food supplies by the middle of the next century. However, this macro data needs to be de-segregated for individual nations to have a realistic estimate of food and population.

However, in the context of developing nations like ours, the ghost of Malthus always haunts us as our population, despite efforts made to control it, is on the rise while our agricultural production and productivity is on the decline. Even our industrial output is not to the desired level and the tertiary service sector of our economy needs to be managed with good governance so as to contain or reduce environmental wastages, hazards and damages.

There is a need not only to check population growth but also accelerate the process of development to sustain our growing population by providing them with minimum facilities, such as nutritious food, shelter and clothing, etc., to protect our large proportion of population affected by poverty. This would warrant us to focus on improving the quality of population. It has been argued that in order to have quality population, there is a need to invest in health and education (Schultz, 1980; World Bank, 1994; Dreze and Sen, 2003). In the context of developing nations, special emphasis has to be laid on the health and education of women in improving the quality of population. The health aspects of population, especially reduction in maternal, infant and child mortality, need to be strengthened by providing a sound reproductive health care system.

It has been observed that population growth takes place within the family or the household level outside the market system, because both the consumption and production side of the analysis needs to be related to the family size and age structure of its members (Ohlin, 1976). Within the utility maximization model, we need to study these demographic variables in interaction with the production and consumption side of the family. It has been suggested that a change in the consumption behaviour of the household or family as a result of rising population size may affect the production behaviour of the family by changing the relationship between labour and capital endowments, between leisure and labour and by affecting the household budget (FAO, 1977). Besides this, the non-economic factors both at the macro and micro-level analysis of determinants of fertility have been observed to be significant in population growth (Birdsall and Faruquee, 1977).

Even Leibenstein (1981) is aware of the limitation of the economic rationality theory on human fertility behaviour because of the difficulty of spelling out the ways in which decisions are made due to lack of information and point of decision procedures to arrive at the maximization. He also points out that without specifying the context of utility, it may be difficult to determine its relevance to the decision-making process. As such, it is necessary to examine the fertility behaviour within social psychological perspectives related to diffusion and adoption of innovations, effective communication models, conscientization methods, group dissonance theory and utility expectancy framework, including reward-punishment, influencing the motivation of individuals Even education through formal or non-formal methods may prove to be a critical factor in influencing fertility behaviour (Mehta, 1997).

As such, in addition to the health aspects of the population, we need to make family planning and information services accessible to all families through population education measures and effective communication pıogrammes so as to promote a small family norm in relevance with the cultural values of people, while respecting the principles of voluntary decisions (Salas, 1983).

In this backdrop, especially in the context of a developing nation like ours, we have to consider environmental programmes or evolve a policy by taking into account both the health and population dimensions of people for sustainable development. There appears interlinkage and permeations of environment, health and population aspects having a bearing on providing quality of life not only to people of the present generation but also to that of future generations. As such, the thrust of the present study will be on these three dimensions: environment, health and population.

References

Beck, U. (1992), *Risk Society*, Beverly Hills, CA: Sage Publications.

Beck, U., A. Giddens and S. Lash (Eds) (1994), *Reflexive Modernization*, Cambridge: Polity Press.

Birdsall, N. and R. Faruquee (1977), *Population, Development Relationships: Approach to Analysis*, Washington, DC: The World Bank.

Black, John (1970), *The Dominion of Man: The Search for Ecological Responsibility*, Edinburgh: Edinburgh University Press, p. 121.

Bookchin, M. (1971), *Post Scarcity Anarchism*, Berkeley: Ramparts Press.

Burch Jr., W.R. and D.R. Field (1988), *Rural Sociology and the Environment*, Westport, CT: Greenwood Press.

Burch Jr., W.R., N.H. Cheek and L. Taylors (Eds) (1972), *Social Behaviour, Natural Resources and the Environment*, New York: Harper & Row.

Burkett, P. (1996), 'Some Common Misconceptions about Nature and Marx's Critique of Political Economy', *Capitalism, Nature Socialism*, Vol. 7, pp. 57–80.

Burton, P.S. (1993), 'Intertemporal Preferences and Intergenerational Considerations in Optimal Resource Harvesting', *Journal of Environmental Economics and Management*, 24 March, pp. 119–132.

Buttel, Frederick H. (2000), 'Classical Theory and Contemporary Environmental Sociology: Some Reflections on the Antecedents and Prospects for Reflexive Modernization Theories in the Study of Environment and Society', in *Environment and Global Modernity*, Gert Spaargaren, Arthur P.J. Mol and Frederick H. Buttel (Eds), London: Sage Publications.

Carson, Rachel (1963), *Silent Spring* (Revised Edition 1970), Harmondsworth: Penguin.

Catton, W.R. and R.E. Dunlap (1978), 'Environmental Sociology: A New Paradigm', *The American Sociologist*, Vol. 13, pp. 41–49.

Catton Jr., W.R. (1980), *Overshoot*, Urbana: University of Illinois Press.

Cernea, Michael M. (1994), 'The Sociologist's Approach to Sustainable Development', in *Making Development Sustainable: From Concepts to Actions*, Ismail Serageldin and Steer Andrew (Eds), Washington, DC: The World Bank.

Commoner, Berry (1963), *The Closing Circle* (Revised Edition 1971), London: Jonathan Cape, p. 188.

Cottrell, Fred (1955), *Energy and Society*, New York: McGraw Hill.

Dickens, Peter (1992), *Nature and Society*, Philadelphia: Temple University Press.

Dreze, Jean and Amartya Sen (2003), *India: Development and Participation*, New Delhi: Oxford University Press.

Dunlap, R.E. and W.R. Catton (1979), 'Environmental Sociology', *Annual Review of Sociology*, Vol. V, pp. 243–273.

——. (1994), 'Struggling with Human Exemptionalism: The Rise, Decline and Revitalization of Environmental Sociology', *The American Sociologist*, Vol. 25, pp. 5–30.

Ehrlich, P. (1972), *Population, Resources, Environment*, San Francisco: W.H. Freeman.

Evans, Peter (1995), *Embedded Autonomy*, Princeton: Princeton University Press.

——. (1996), 'Government Action, Social Capital and Development: Reviewing the Evidence on Synergy', *World Development*, Vol. 24, pp. 1119–1132.

——. (1998), *Liveable Cities?* Unpublished Manuscript, University of California, Berkeley: Department of Sociology.

Firey, W. (1960), *Man, Mind and Land*, New York, Free Press.

Food and Agriculture Organization (1977), *Population and Agriculture Development*, Rome.

Giddens, Anthony (1987), *Social Theory and Modern Sociology*, Stanford: Stanford University Press.

Goldblatt, D. (1996), *Social Theory and the Environment*, Boulder, CO Westview Press.

Harris, M. (1978), *Cannibals and Kings: The Origins of Cultures* London: Fontana.

Harvey, Brian and John D. Hallett (1977), *Environment and Sociology An Introductory Analysis*, London: The Macmillan Press.

Holling, C.S. (1996), 'Engineering Resilience versus Ecological Resilience', in *Engineering with Ecological Constraints*, P.C. Schulze (Ed.), Washington, DC: National Academy Press.

Leibenstein, Harvey (1981), 'Economic Decision Theory and Fertility Behaviour', *Population and Development Review*, Vol. 7, No. 3.

Lender, Warren (1989), 'Preface to Special Issue', *Development*, Vol. 2, No. 3 (Sustainable Development: From Theory to Practice), Rome: Society for International Development.

Malthus, Robert (1798), 'On the Principle of Population', as quoted in *Environment and Society: An Introductory Analysis* by Brian Harvey and John D. Hallett, London: The Macmillan Press, 1977.

Marsh, George Perkins (1864), 'Man and Nature or Physical Geography as Modified by Human Action', as quoted in *Environment and Society: An Introductory Analysis* by Brian Harvey and John D. Hallett, London: The Macmillan Press, 1977.

Maxneef, A. Manfred (1990), 'Notes on the Semiology of Our Mega Crisis', *Development*, Vols 3 and 4 (Human Centered Economics: Environmental Global Sustainability), Rome: Society for International Development.

Meadows, D. et al. (1972), *The Limits to Growth*, A Report for the Club of Rome, East Island, New York.

Mehta, S.R. (Ed.) (1997), *Poverty, Population and Sustainable Development: Essays in Honour of Professor V.S. D'Souza*, Jaipur and New Delhi: Rawat Publications.

Murphy, R. (1994), *Rationality and Nature*, Boulder, CO: Westview Press.

Nelissen, N.J.M. (1972), *Sociale Ecologie*, Utrecht/Antwerpen: Het Spectrum.

Norton, B.G. (1955), 'Future Generations, Obligations to', *Encyclopedia of Bioethics* (Second Edition), New York: Macmillan.

——. (2003), *Searching for Sustainability: Interdisciplinary Essays in the Philosophy of Conservation Biology* (Cambridge Studies in Philosophy and Biology), Cambridge: Cambridge University Press.

Norton, B.G. and Michael A. Toman (1997), 'Sustainability: Ecological and Economic Perspectives', *Land Economics*, Vol. 74, No. 4.

Ohlin, G. (1976), "Economic Theory Confronts Population Growth" in Ansley Coale (Ed.), *Economic Factors and Population Growth*, New York, John Wiley and Sons.

Osborn, Fairfield (1948), 'Our Plundered Planet', as quoted in *Environment and Society: An Introductory Analysis* by Brian Harvey and John D. Hallett, London: The Macmillan Press, 1977.

——. (1953), 'The Limits to the Earth', as quoted in *Environment and Society: An Introductory Analysis* by Brian Harvey and John D. Hallett, London: The Macmillan Press, 1977.

Parsons, H.L. (Ed.) (1977), *Marx and Engels on Ecology*, Westport, CT: Greenwood Press.

Peirce, C.S. (1960), 'Elements of Logic', in *Collected Papers of Charles Sanders Peirce* by C. Hartshorne and P. Weiss (Eds), Cambridge, MA: Harvard University Press, as quoted in B.G. Norton, *Searching for Sustainability: Interdisciplinary Essays in the Philosophy of Conservation Biology* (Cambridge Studies in Philosophy and Biology), Cambridge: Cambridge University Press, 2003, pp. 88–103.

Salas, M. Rafael (1983), 'The 1983 State of the World Population', Report, United Nations Fund for Population Activities, Excerpts 'The Family of the Future', *Development Forum*, Vol. XI, No. 6, pp. 8–9.

Salim, Esmail (1991), 'Towards a Sustainable Future', *Development*, Vol. 2 (Vision Fused with Urgency: A New International Cooperation), Rome: Society for International Development.

Schnaiberg, A. (1980), *The Environment*, New York: Oxford University Press.

Schnaiberg, A. and K.A. Gould (1993), *Environment and Society*, New York: St. Martin's Press.

Schultz, W.T. (1980), *Investing in People: The Economics of Population Quality*, Delhi: Hindustan Publishing Corporation.

Schumacher, E.F. (1973), 'Modern Pressures and the Environment', *Christian Action Journal*, Summer.

Schwartz, E.S. (1971), *Overskill: The Decline of Technology in Modern Civilization*, Chicago: Quadrangle Books, p. IX.

Serageldin, Ismail (1994), 'Making Development Sustainable', in *Making Development Sustainable: From Concepts to Actions*, Ismail Serageldin and Andrew Steer (Eds), Washington, DC: The World Bank, pp. 1–5.

Solow, R.M. (1993), 'Sustainability: An Economist's Perspective', in *Economics of the Environment: Selected Readings*, Robert Dorfman and Nancy Dorfman (Eds), New York: W.W. Norton and Company.

Spaargaren, Gert, Arthur P.J. Mol and Frederick H. Buttel (2000), *Environment and Global Modernity*, London: Sage Publications.

Vogt, William (1948), 'The Road to Survival', as quoted in *Environment and Society: An Introductory Analysis* by Brian Harvey and John D. Hallett, London: The Macmillan Press, 1977, p. 65.

Ward, Barbara and Rene Dubos (1972), *Only One Earth*, New York: Norton.

Webb, W.P. (1952), *The Great Frontier*. Boston: Henghton Mifflin.

Weber, M. (1968), *Economy and Society*. Berkeley: University of California Press.

West, P.C. (1984), Max Weber's human ecology of historical societies, in V. Murvar (Ed.) *Theory of Liberty, Legitimacy and Power*. Boston: Routledge and Kegan Paul, pp. 216–234.

Wilkinson, R.G. (1973), *Poverty and Progress*. New York: Praeger.

World Bank (1994), *World Development Report*, New York: Oxford University Press.

2

Development and Fragile Ecological Environment

It has now been well recognized that the environmental crisis has been mainly caused by rapid expansion of industrial development for economic growth due to over-exploitation of natural resources.

In the present chapter, an attempt will be made to highlight the economic development and the consequent environmental destruction, damage and hazards caused on planet earth for providing a comfortable social life to people inhabiting it. Besides, in acquiring sustainable development, the moral aspects related to human welfare will be under focus as part of environmental ethics and philosophy. Further, human welfare aspects within the ecosystem health in the context of sustainable development shall also be highlighted. This will be followed by environmental challenges in the era of globalization and modernization.

Economic Development and the Environmental Crisis

In conceiving an environmental crisis in terms of environmental destruction, damage and hazards, one can consider it in distinct geographical regions of the planet. In the Northern Hemisphere, countries such as erstwhile USSR (present Russia), Europe, North America and Japan, especially after the end of the Second World War, had suffered ecological disruptions mainly through the pollution problems arising out of rapid and often uncontrolled industrial development. On the other hand, ecological disruptions in the Southern countries have been caused mainly due to the pollution of the land, air and water, because of economic underdevelopment as a consequence of lack of consumer demands, food and hunger problems and rapid population growth.

These developing nations, earlier known as Third World nations (Africa, Asia and Latin America), were having dependency relationship with the major industrial powers due to the extent of poverty, hunger, population growth rates and economic stagnation. Most of the nations in the Southern Hemisphere were either colonies or dependent economies of the developed nations. It is observed that there is some linkage between the particular forms of environmental disturbance in the overdeveloped and in the underdeveloped world. In that link, perhaps there is an operation of the world market or free enterprise system. In fact, the capitalist mode of production appears to be central in understanding the nature of the existing global environmental crisis because the mode of production generally gives an idea about a nation's social structure and the development of its external relationship with the other countries. It is argued by some scholars that under capitalism the economic system begins to dominate the rest of the society rather than getting submerged in it, as the earlier postulate of an individual's 'just' share of the community 's collective material resources is replaced by the possibility of individual starvation amid group plenty (Reich et al., 1972).

As such, it is the dependency relationship that exists between the rich nations and the poor nations that is the root cause of most of today's environmental problems. The economic production unit under this capitalist system has been employed to generate surplus of goods and services to promote the growth of the productive enterprise and to provide work for the people and raise its external power vis-à-vis the other nations. The surplus accumulation by the rich nations has been mostly appropriated through surplus value from labour, primitive accumulation through outright theft or exploitation of the wealth of another country. Further, it is accumulation through the development of external colonies by exploiting the people and natural resources of the poorer countries that has led to the development of surplus capital by the rich nations for their domestic advancements. In short, it is through exploiting the skills and labour of slaves brought into the home country to benefit the internal economy of a nation as a form of cheap labour and accumulation through trade surplus by selling products, goods and services than purchasing from abroad that a balance of payment surplus has been generated in order to reinvest the same in the internal economy to gain even more capital formation (Miller, 1996).

Most of the developing countries still look forward to the first, that is industrialized, rather than the second (socialist) world such as People's Republic of China for the containment of pollution due to industrialization. It is observed that the economic development of the developing nations remains peripheral to the developed nations mainly because of massive accumulation of economic surplus by the internal industrial development of Europe and North America. This dependency syndrome of the developing nations on the rich developed nations has serious implications for the environmental crisis caused by hunger and starvation, rapid population growth and poverty, etc. To tackle these environmental issues, there is an urgent need to restructure the global market economy as without that, there is not much hope to resolve the pollution problems of the overdeveloped world and the macro environmental issues confronted by the poor nations (ibid.).

Notwithstanding the above observations made by neo-Marxists' school of thought which advocates the dependency theory and world system theory in response to modernization paradigm emphasizing industrialization, it is also noted that the import substitution syndrome for industrialization of the developing nations has also caused environmental degradation because of lack of adequate scientific knowledge and non-availability of adequate scientific manpower and clean technology with them. It has been emphasized by the development theorists that the development of developing nations through the transfer of technology from the developed world may not be a necessary condition. In fact, there is a need to make development of a nation relevant with the cultural needs and aspirations of the people inhabiting it.

Nature provides a suitable ecological balance and a natural process of recharging the resources in due course of time. Till man's intervention in nature remained limited, this remained in perfect order. However, with the increasing population and consequent rise in man's needs and wants, the resource utilization far exceeded the recharging capacity of nature and this led to environmental degradation. There is also a need to develop indigeneous technology by the developing nations, such as intermediate technology or alternative technology, instead of banking heavily upon the highly sophisticated and complex technology of the Western world. They could make it possible to have it in view of the rich inputs in research and development which we the poor nations cannot afford to invest in to the extent they could do it.

There is a need to have a healthy environment for each nation, developed or developing, in which one can have the highest quality of life with minimum of environmental problems. In developing nations like ours, we have gone in for industrialization and subsequent urbanization by adopting the technological inputs of the West and this has led to pollution of the environment through industrial discharges, effluents, poor quality of air, water, insanitary living conditions, problems of solid waste disposal and above all, the degradation of natural resource base, both in the rural and urban social life. For example, it has been reported that the air pollutants not only have harmful effects on human health but they also influence crop yields, natural resource base and property. It has been observed that the acid rain, which is due to air pollutants, can damage forests, crops, property and human health.

In the contemporary modern scenario, the air pollution is mainly caused by industrial and automobile emissions as green house gases such as methane and carbon dioxide. These are released by them and lead to global warming, which may result in depletion of the ozone layer to make living uncomfortable. The developing nations in order to check air pollution must utilize or evolve cleaner automobile technologies in the urban areas, while in rural India, there is a need to have smokeless or biogas-based *chullhas* so as to reduce the incidence of air pollution. There is also a need to have an efficient public transport system in the congested mega cities as well as in the upcoming towns so as to cut down on the density of air pollution because of increase in the volume of personal vehicles.

Water is another drink provided by nature which is necessary for sustenance of life but when it gets contaminated with pollutants, it causes many kinds of health hazards. The main sources of water contamination are uncontrolled disposal of urban waste, disease-bearing human waste, industrial waste flowing into water bodies, toxic chemicals and metals, as a result of agricultural operations and sewage disposal.

In addition to the air and water pollution, solid waste is another major source of providing environmental hazards in a developing nation like ours. Apart from evolving environment-friendly technologies for its treatment to reduce the quantum of waste and to convert it as a source of energy and compost, there is also an urgent need to adopt improved methods of its collection, transportation and treatment. There has to be separate dumping of recyclable and non-recyclable waste, besides imposing a ban on the use of polythene bags, treatment of garbage dump sites, separate efficient and effective treatment of hospital waste and health care for rag pickers (Chandna, 1998).

In view of a crunch of financial resources in developing countries like ours, though attempts are made by the nation-state to introduce some of these technologically-oriented procedures and methods along with legislative measures, the implementation at the ground level falls short of the expectation of the people. This is the main reason for the causation of various kinds of environmental degradation and health hazards not only for the present, but also for the future generations.

As an agricultural society, we also face soil erosion due to land degradation as a result of weathering of land surface by running water, wind, waves and moving ice because of long and continuous agricultural operations of the land to produce crops. It has also been realized that the application of chemical fertilizers to the soil to improve the productivity has also resulted in soil pollution due to the impact of fertilizers upon the physical, chemical and biological properties of the soil. In addition to heavy inputs of fertilizers causing poor soil health, we have observed problems related to waterlogging, salinization and laterization, ultimately resulting in desertification of the soil. No doubt, the inappropriate quantum of use of fertilizer and irrigation inputs has been observed as causing soil pollution, but both these technologies have been employed by the land cultivators to increase the productivity of the soil. However, agricultural scientists suggest crop rotation, strip cropping, growing seasonal crops as a cover and application of green manure to conserve the soil as part of the vegetation management measure.

The other two important resources that have been exploited for economic development are forest covers and wildlife. Man, in order to have comfortable and good living conditions, has exploited all types of wooden areas such as natural forests, woodlands and shrubs and plantation of trees, etc. These forest covers are desirable as natural checks to prevent soil erosion and siltation of lands, supplying timber for various uses, supplying raw material for forest-based industry and to provide a hunting ground for hunters as excellent recreation sites.

It is estimated that nearly 40 per cent of the total land area of earth at present is covered with forests. Out of these, nearly two-thirds area constitutes natural forests while one-third area is covered by plantations, woodlands and shrubs. It is the growing population in all parts of the world that has posed serious threats to the forest cover. By 1800 AD, the population of the world was around one billion while at present it hovers around 6.3 billion. This has caused great pressure on the forest cover. We have roughly 40 per cent of the world's total area which has been brought under crops. It is observed that during 1960–90, nearly 20 per cent of the tropical forests in the developing nations have been cleared and nearly 50 per cent of the trees have been axed and are being used as fuel irrespective of their timber value because of increasing requirement of land for both the fuel wood and crops. As such, the world's forests are being cleared not only for commercial purposes but also for fuel wood and agricultural purposes (ibid.).

Acid rain is considered to be a dangerous pollutant which gets introduced in the atmosphere because of industrial activities. Sulphuric acid as well as nitric acid fall on the ground and on trees as acid rain. Besides, the vegetation absorbs sulphur dioxide (SO_2) gas directly from the atmosphere. Once acid rain occurs, it can lead to a chain of chemical reactions, which ultimately culminate in forest death. The forest degradation due to acid rain in the highly industrialized countries of the temperate regions is being prevented on a priority basis by controlling the emissions of harmful chemicals of fossil fuels' power generating industries, and by going in for industries having low emission level of sulphur and other chemicals and through use of alternative technologies such as atmospheric fluidized combustion, new filtration process in Germany and 'sea water technology' in UK.

Deforestation is on the rise in the Central American countries of Haiti, Jamaica, Cuba and Barbados, which seem to have lost 66 per cent of their natural forest cover by now. In tropical countries, the minimum level of forest requirement is 18 per cent, while India has only 10–12 per cent of the area under forest. This needs to be raised to 18 per cent. The main reasons identified for causing deforestation are rapid growth of population resulting in settlement of new areas, extension of agriculture requiring forest clearing, shifting cultivation as practised by many tribals, poverty of the people compelling them to clear forests for fuel wood, development of transport network such as roads, railways, airways, construction of dams, growing industries and housing projects demanding more of timber and pulp, etc., forest fires causing destruction of forests, acid rain which can cause forest death and compulsive administrative decisions.

All these factors responsible for deforestation have serious consequences as they lead to extensive soil erosion, occurrence of floods, loss of certain species, climate changes, difficulty in hydrological operations because of siltation and water flow, various kinds of impact upon economic health, significant impact upon human health and the loss of aesthetics. Both afforestation, that is forestation of new area, and reforestation of the deforested area or regeneration of the forest in the deforested area are being recommended for conservation of forests, along with promotion of forestation in areas unsuitable for cropping, agro-forestry, social forestry, sustainable forest management and education of people to build a people's movement against destructive use of forests by the environmentalists. For example, the World Rain Forest Movement at the international level has called for banning of aid for such development projects which are taken up in the forested areas as well as to ban the trade in timber from virgin forest areas. In India, the Chipko movement started by Gandhian stalwart Sunder Lal Bahuguna in Uttarakhand in 1972 to protect the forests has now spread to many states, including Himachal Pradesh, Jammu & Kashmir, Rajasthan, Arunachal Pradesh, Karnataka, Bihar, Madhya Pradesh, Maharashtra, etc. This movement has been successful in banning of tree felling in all areas having a slope of more than 30 degrees and an altitude of 1,000 metres above sea level. Even social forestry or community forestry has involved farmers and local people to meet the forest needs of the household and community. Social forestry through planting trees having household utility has been very successful in the states of Gujarat and West Bengal by bringing marginal land under forests as well as in providing employment or subsistence livelihood needs to many (ibid.).

Wildlife is another natural resource that has been subjected to exploitation by mankind for its advantage and pleasure-seeking adventures. Wildlife in a generic sense includes all forms of life, that is the microbes, plants and animals with the exception of cultivated plants and domesticated animals. It is difficult to make an estimate about the actual number of wildlife species due to numerous varieties of species of plants or animals. In the Indian scriptures, 84 lakh *yonis* (forms of life) have been indicated.

India is having a great variety of climatic conditions which help in sustaining many diverse forms of life. Though India forms only 2 per cent of the world's total land surface, it supports more than 5 per cent of the world's fauna and flora, including some strains of African, European as well as South-East Asian biological systems, thereby presenting a large variety and richness of wildlife. It has been observed that despite conservation measures, certain species of wildlife are getting extinct mainly because of destruction of their natural habitat, excessive hunting and weak law-enforcing agencies. We need to sustain a large variety of native wildlife because of ecological, biological, scientific, economic and ethical requirements.

In order to ensure preservation and to sustain all varieties of wildlife in their natural habitats, a wildlife management system needs to be evolved particularly for the game animals; we need to adopt an approach that should provide legal protection, habitat preservation, assemblage protection and species preservation.

In addition to adopting various strategies for the conservation of wildlife such as preserving the natural habitat, basic life support systems of soil, water, air and energy, prevention of overexploited use of ecosystems to ensure sustainability of the habitat and enacting national and international laws to ban trade in endangered species and hunting to threatened species, setting up of national parks, sanctuaries, zoological parks and rehabilitation centres, there is an urgent need to create environmental awareness among public in general and school and college children in particular about wildlife preservation measures. India enacted the Wildlife Protection Act in 1972, which has provisions for imposing penalties to those who commit offences related to wildlife preservation and protection. In our Constitution, there is a provision which directs the state governments to protect and improve the environment and to safeguard the forests and the wildlife so as to minimize environmental degradation, damage and hazards (ibid.).

An overview of exploitation of natural resources by man to his advantage for his comfortable living to pursue economic development of the group, community or a society, of which he forms a part, has amply demonstrated that the excessive use of natural resources such as air, water, soil, forest covers, wildlife, etc., has resulted in having a fragile ecological environment on earth. As humans, we need to show commitment to strike a balance between the environment and development not only for a good life for us, but also for our future generations as the natural resources on the planet are finite and are depleting as we move on to the next generation.

Energy as Source of Pollution

Rapid industrialization and continuous growth of population have necessitated an increase in the use of energy. It has caused environmental pollution as no energy resource is considered completely environmentally safe. There is a need to optimally utilize energy so as to minimize the environmental hazards. No doubt, solar energy is the main source for the existence of life, yet the real energy resource revolution started with the efforts of man to harness fire for warmth, light and protection. Man also evolved environmentally-friendly alternatives through conventional use of petrol and diesel. Besides, the excessive exploitation of fossil fuels (such as coal and oil) has generated green house gases, causing acid rain.

The bio fuels such as petrol and diesel are yet to be put to commercial use and it may be possible to use them effectively for the production of energy by the middle of the 21st century or till such time when the pressure on fossil fuel reserves compels us to make a shift because of their depletion or diminishing of quantities. Besides exploiting the concentrated energy in coal, oil and natural gas, man has the ability to also organize a natural process of photosynthesis in the form of agriculture as well as to tap the wind energy through wind mills.

There are a number of sources of energy which have been employed by mankind, including fossil fuels (coal, oil, natural gas), nuclear power, hydro-power, wind power, tidal energy, solar energy, geo-thermal energy, and biomass energy in the form of wood, dung, agricultural waste/crop residue. It is observed that in less developed countries, nearly 66 per cent of the energy requirements are met by wood, dung, agricultural waste and chemical energy in food. These sources account for nearly 15–20 per cent of the total energy used.

About a 100 years back, say around 1880, these sources provided more than 50 per cent of the total energy used in the world but with the growing population pressure, the demand for energy has also risen and it has grown at a rate faster than population (for example, population growth in the world had its peak of 2% per annum during 1950–1980 but the rate of growth of energy for this period was 3.5% per annum).

Presently, the use of energy is growing much faster in the less developed countries because of their expansion in the industrial sector and mechanization of the agricultural sector. As such, it is a big challenge for them to evolve cleaner technologies for tapping their energy resources which emit less smoke while making a shift from wood and dung, etc. It is paradoxical that the more developed countries have nearly 25 per cent population but have energy consumption of 83 per cent, while less developed countries have about 75 per cent of the population but 17 per cent energy consumption. It is also reported that Asia has the largest reserves of coal, oil and one-third of the natural gas reserves of the world, while other continents together have only half of Asia's oil reserves. North America ranks high in terms of uranium, while Europe ranks high in terms of hydro-power potential. Europe is reported to be the largest consumer of energy (33% in the world, followed by North and Central America, Asia, South America, Africa and Oceania in this order). Between 1973 and 1993, the energy consumption increased in the world by 47 per cent. The highest increase in consumption was observed in Asia, followed by Africa, South America, Oceania, North and Central America in this order. In India, the energy consumption during this period increased by 258 per cent (ibid.).

Every energy, whether conventional or non-conventional, renewable or non-renewable, has some environmental-degrading effect. Coal is considered to be largest conventional fossil fuel resource because mining of coal is not only health hazardous for its workers, but it also emits a particular matter, sulphur dioxide and carbon dioxide into the environment. No doubt, cleaner technologies have now been evolved to reduce the emission level from coal use (for example, gas desulphurization, integrated gasification, hybrid combined cycle system and more latest, combined coal gasifiers and cells and magnet-hydrodynamics). Oil, on the other hand, is less a generator of carbon dioxide and sulphur dioxide in comparison to coal but its combustion also releases carbon dioxide and sulphur dioxide, causing serious environmental pollution, including acid rain, but its main problem is oil import dependence and oil import vulnerability.

Natural gas is comparatively considered cleaner than coal and oil, but the leakage of methane gas from gas distribution pipelines and green house gases from oil and gas wells has contributed to global warming. Nuclear energy was considered a very apt modern answer to world's future energy requirements, but we had nuclear accidents at Three Mile Island (USA) and Chernobyl (Russia), which released radioactive energy material as a part of nuclear waste disposal. This may be due to some human error.

The long life and high toxicity of radioactive chemicals, besides transportations of nuclear base, have resulted in enlarging the magnitude of risks involved. It is reported that due to the high risk involved, Sweden which was having twelve nuclear power plants started de-commissioning their power stations. Hydrogen is also suitable for generation of chemical fuels as well as for indirect utilization of solar energy, but there are technological limitations of preserving hydrogen and oxygen in a combined form to form water vapours, when these gases cool down. This area needs to be researched further.

In the context of developing nations, energy from fossil fuels and nuclear source is essential for the growth and development, but the renewable sources of energy such as hydro power, wind power, tidal wave power, geo-thermal energy, solar energy and bio-mass energy need to be exploited more as they are cleaner and are non-contributory to acid rain. They are also more diversified and provide greater energy security. For example, in India, many coastal states like Kerala and Gujarat have established huge wind mill farms but even there, these are not without environmental hazards such as noise pollution, electro-magnetic interference and ugly sites.

In our context, we cannot ignore the energy potential of bio-mass fuels, from plant and animal matter waste and residues. The land-filled waste with household waste having a temperature of 35 degree Centigrade is ideal for bio-gas generation. It can also be used to generate heat and electricity. If bio-mass is not utilized for energy generation, it can release green house gases, methane and carbcn dioxide into atmosphere, resulting in global warming and ozone layer depletion. This environment-friendly energy source needs to be optimally utilized and subjected to technological innovations such as liquid bio-fuels like bio-diesel and bio-ethanol as more environment-friendly alternatives than the conventional fossil-based fuels, in order to have minimum adverse effect on the environment (ibid.).

An account of different energy resources given in the preceding pages as a part of industrialization processes and mechanization of agriculture in the developing nations help us to further examine the environmental challenges in the era of globalization and modernization.

Environmental Challenges in the Globalized Economy

In the context of resources and ecological sustainability, there are three types of relationships between economic growth and sustainability that need to be highlighted. First, it is argued that sustainability and economic growth can be compatible if we recognize the need for minimal international regulation to protect endangered ecosystems and species. The second view is opposite to it. This emphasizes that economic growth means increased utilization of energy and material in an economy that may undermine the sustainability of the environment. The third perspective focuses that economic growth whether or not compatible with sustainability will depend upon prior definition of concepts such as wealth, the interest of future generation and the nature of economic efficiency. The third perspective is worth consideration as a priority of sustainability, that is, as a goal rather than as a set of expost management tools.

The problem of climate change as a consequence of spread of pollution around the globe requires some analysis. It is observed that in the rural and semi-urban areas, indoor pollution from burning wood and poor ventilation, harms more than four million people world wide and leads to acute respiratory infection resulting into death of four million children annually. The household sewage is another global problem which has its beginning in the family. It is the major cause of water contamination and poor sanitation facilities for more than one and half billion people globally and results in even larger number of infant deaths. In the context of developing nations, it is reported that nearly 95 per cent of sewage is untreated. The clean drinking water if provided may result in cutting down the death of two million children each year from diarrhoea. It is estimated that nearly 13.5 million children die due to combined effects of poor diet and domestic insanitation (Ekins, 1996, as quoted by Redclift, 2000).

It is also observed that during 1980s, the number of people without adequate sanitation facilities increased many-fold in various parts of the world because of growing water scarcity. Further, nearly 40 per cent of the global population faces periodic droughts and there has been a rise in the per capita water consumption for irrigation in agriculture at the cost of domestic water usage. The other problem faced by cities in the world is that of quality of the air. Nearly a billion-and-a-half people live in the cities with air quality below minimum standards laid down by the World Health Organization. This has caused respiratory and other ailments from which half a million people die prematurely every year (ibid.).

Further, the production of industrial waste and the extinction of species are not only linked to levels of consumption, but also to resource degradation through unsustainable economic growth as that puts individual livelihood at risk throughout the developing world. These are, in fact, survival issues for the poor and a matter of concern for policy analysis. Globalization has not only to tackle the global environmental problems of environmental change, but also the issues of direct survival of the poor. Globalization has to set a uniform pattern of economic goals through competitive processes to achieve them. It has, therefore, to safeguard the survival of local and frequently the opposite cultures. In the global environmental problems as a result of integrated capitalist market economy, there is diffusion of sources and that relocates waste and pollution globally (for example, spread of polluting industries and energy and transport technologies from the developed to the developing nations). Further, there is diffusion of impacts through the medium of water, soils and the atmosphere, resulting in global warming, ozone depletion and acid deposition. These sets of problems are not only a result of development of the global economic system but are also embedded in the conventional societal and socio-cultural processes. In this context, Giddens' (1994) concept of globalization in terms of intensification of worldwide social relations, linking distant localities in such a manner that local happenings get shaped by events furthering many miles away and vice-versa, appears appropriate in the sense that the sight of environmental degradation gets removed from its agent of causation.

In the globalized economy, it is observed that the location of material sources and the sites where these are transformed and where they are consumed and the location of waste things invariably do not coincide because of interdependent patterns of development based on the mobility of capital and the relocation of industrial processes through the international division of labour by MNCs having comparative advantage and regional patterns of specialization. In this regard, a right question posed by Escobar (1996) was to consider the visualization of Earth as a fragile ball by making a reference to the *Scientific American* (September 1989) special issue on 'Managing the Planet Earth'. This focused on 'what kind of planet do we want, what kind of planet can we get'.

It is observed that the present international political economy is managed by a small number of multilateral institutions, mainly the World Bank, the International Monetary Fund and the World Trade Organization, having policies determined by the richest industrialized nations and Group of Seven (G-7), which control over 60 per cent of the world economic output and over 75 per cent of world trade. Further, the environmental effects of economic globalization have been observed to be uneven, both spatially and socially. The diffusions of sources have been determined by the economic development processes through investment, trade, etc., by transnational companies, multi-lateral institutions and the richest industrial nation-states while the diffusion of system influences have been observed to be more diffused invariably as irreversible and likely to be passed on to future generations (Redclift, 2000).

In the era of globalization and modernization, the developing nations have to bear the brunt of environmental problems as the developed nations appropriate both the industrial and agricultural sectors' economy through technological inputs and mechanized agricultural farming operations, through agents and agencies of industrialized rich nations and multilateral institutions. This marginalizes the poor developing nations at the cost of the affluent nations.

In the wake of above, the developed nations must address their primary responsibility for the present environmental crisis and make efforts to redress it, as the present structures of interdependence of trade, aid and debt make sustainable development a difficult proposition for the developing nations. The poor in these nations are the worst sufferers and they should be considered as the main target for sustainable development. There is a need to provide equitable resources to them through grassroots movements. Further, the developed nations must be liberal towards the developing nations to get them out of the debt traps and provide fair trading relationships in global economic institutions. Besides, the developed nations should be ready to provide concessional transfer of clean and efficient technologies for suitable industrialization and environmental regeneration of the resources of the developing nations. It is argued that the return of illegal flight of money from the developing to the developed nations, effective land reform programmes by granting peasants indigeneous land rights and a movement towards greater democracy to make people masters of their own development are some of the necessary requirements in the developing nations for sustainable development (Ekins, 1991).

But the moot question is whether or not the developed world will provide efficient and clean technology to the developing world and if they do so, at what cost? Should we depend upon the transfer of technology from the West or evolve our own indigenous technology for protection and regeneration of our environment? Further, should we pin hope on the technological innovations of the scientists to optimally use the existing resources and improve the quality of life for all on this planet? The most significant question pertains to the fact whether or not the unequal partners in the global sustainable development will come forward to negotiate on equal terms. In this asymmetrical globalized world economy and the contrasting living conditions and lifestyles of people between the developed and developing nations, one has to address the importance of equality to perceive and tackle environmental problems. In the context of the poorest countries, the environmental problems are associated with health, shelter and food availability. On the other hand, in the case of newly industrializing nations, environment is to deal with the short-term and medium-term cost of following very rapid economic growth such as high levels of air pollution in the cities, whereas in the developed world environment is increasingly getting involved in highly invisible and unforeseen risks such as levels of radioactive and chemical exposure. In this regard, the issues of equality are less important while tackling these problems than freedom of information or civil rights.

There is no doubt that environmental issues are subject to ideology in terms of behaviour of others towards the environment and this differs in different societies and cultures. This apart, the case of climate change also poses the problem of equity in owning up to the global environmental problems between the developed and the developing nations. Because of the variations observed in the calculations of atmospheric emissions, major economic agencies of the developed nations do not encourage the developing nations to shift to cleaner technology in order to maintain market hegemony over the labour. It has a displacing effect of hi-tech industries being recommended to countries where labour costs are high.

In the above scenario, it is very difficult to conceive of both the intra-generational and inter-generational equity for the developing nations as they are guided more to reduce inequalities in the present global system than to make sacrifices to have gains for the future affluent nations of the North. In this context, the distributive problem lies in the failure to take global action for climate change because of the different historical reasons for the causation of the same by the developed and developing nations. Further, there is a controversial issue whether or not all have an equal stake in the global environmental problems.

In short, if we place the globalized world economy in the context of a social theory to examine the material forms of globalization and their cultural constructions, including cultural opposition, we can discuss the environmental issues at three levels. First, these can be addressed through different and frequently competing policy frameworks such as market and regulatory mechanism. Second, there is opening up of global markets, global consumption and penetration of remote ecological systems by global capital, but this is in conflict with the developing economies which focus on everyday life of the people. This causes tension for sustainability in the long-term livelihood of people in the developing nations and perhaps can be tackled through global environmental management. However, this remains elusive as it contradicts the integrity of culture and place of the people. As such, the developing countries may show resistance to global environmental management and prefer local-based policy options. Third, the global environmental problems may be addressed through legitimate political actions and a domestic policy which should support global environmental management. However, questions have been raised about the legitimacy of both environmental conditionality and the coercive factors used by global institutions suggesting management solutions. The coercive approach appears to have moral implications as in this case, the integrity of the biosphere is used to justify the creation of a new regulatory world order. It is observed that the global environment poses both a crisis of authority and non-availability of social and political legitimacy. The problem of globalization and environment change needs to be looked into as a form of new configuration as outlined above (Redclift, 2000).

Environmental Philosophy or Ethics

In the wake of material prosperity which has enriched the lifestyle of majority of the people of the developed nations and has also started penetrating into the lifestyle of people of the developing nations, especially in the metros and big cities, it becomes imperative to show concern for the large majority of population who are not only marginalized but pauperized economically, politically and socially across the globe, but much more so in the developing nations. As human beings, it is our moral responsibility to uphold altruistic values and curb the mercenary tendencies or orientations to tackle environmental problems on an equitable basis so that the pains of the poor are reduced in comparison to the gains of the rich for balanced growth and sustainable development.

Attempts have been made by environmental philosophers by evolving environmental ethics to search for justification of general moral principles such as bio-centrism or eco-centrism, that may guide the process of environmental policy development to be more pluralistic and problem-oriented rather than basing it on abstractions and arguments having universal principles. Philosophical ethics involve a variety of conceptualizations of moral obligation. Broadly, three types of general theories are being recognized. First, teleology, which includes utilitarianism of various sorts, welfare economics and various interest theories. The utilitarians have emphasized that all generations do have a chance to maximize their welfare as individuals. This can be measured in terms that are spread across persons and across time by taking into account the intergenerational opportunities and by having rules that may guarantee the future generations to have the opportunity to enjoy a standard of living in comparison to the preceding generations. Secondly, there is a 'rights' theory which focusses on the fact that future people have 'rights' which we must respect. It is argued that if future persons really have 'rights', then these 'rights' would win over any kind of mere enjoyments that the present gains at the expense of future well-being. This is based on the assumption that the time at which a harm occurs is irrelevant to its moral status in the sense that if the decisions and actions of the present strongly damage the life prospects of future generations in avoidable ways, then the present generation will feel guilty of intergenerational damages. Thirdly, there are contractarian theories that have emphasized the significance of at least some 'implicit consent' which may provide an acceptable rate of exploitation and savings in every generation when applied intergenerationally.

Notwithstanding the differences in these approaches, one thing that clearly emerges is that our ability to affect the way of future will impose responsibility that involves issues of justice, fairness and equity and not just questions of economic efficiency. This raises certain environmental policy formulations as to whether we should consider threats to the basic rights of future people or evolve a fair investment policy based on a welfare analysis. Further, can we pursue policies in response to threats to both rights and basic welfare or follow them simultaneously, which may lead to contradictory goals, or have different priorities in protecting the resources for the future? This is nothing but moral monism as it either implies a welfare criterion or rights-based criterion but on the other hand, the pluralistic approaches to ethics recognize multiple principles and criteria by scholars like Stone (1987). He has advocated that 'right' management of the environment will require the application of different principles in various situations. It has been argued that in having environmental policy, there is a need to have a pluralistic system of decision-making by using an efficiency criterion and a conservation criterion as it will have a built-in ecologically-based concept of adaptive management (Norton and Toman, 1997 reproduced in Norton, 2003).

In striking a balance between the material and moral dimensions of environmental issues and crisis, there is a need to follow the Gandhian principle, which advocates need rather than greed of mankind for sustainable development of a society. Even Buddhism preaches that we need not be doomed to the perpetual rat race, which can be broken through wisdom, meditation and morality. As human beings, we need to rise above ourselves and show concern for the welfare of others. This can be achieved through love, cooperation, friendship and goodness among people by inculcating these altruistic values to make environmental ethics a reality as a part of environmental philosophy.

Health of Human Welfare Ecosystem in the Context of Sustainable Development

Sustainability has become a common denominator both among the environmentalists and development thinkers. The term was popularized in the first instance in the context of resource goods, that is in terms of the highest level of exploitation of resources from a forest or fishery while consistently maintaining a steady flow of them. Currently, it is being employed to indicate the levels of pollution and degradation of natural systems, while consistently maintaining a balance between the current levels of use and enjoyment of these systems. Broadly, when we speak of sustainability, we tend to exclude the moralistic dimensions suggested by extreme positions taken by ecologists. Their argument is that the natural world resources should not be considered for human use at all. On the other hand, there is another position taken by those who advocate unlimited economic growth and are in favour of not constraining the ability of the free market to generate goods and services in response to consumer demands. As such, there should not be any constraint on the use of nature for development purposes. Between the two extreme positions, a vast majority of environmentalists believe that though the use of environment is morally acceptable, we should be under obligation not to misuse the environment in unsustainable ways.

The Brundtland Commission's report, *Our Common Future* (1987), has defined sustainable development in terms of the needs of the present without compromising the ability of future generations to meet their own needs. There are two pre-concepts involved in this definition. When we speak of the concept of needs, we need to keep in consideration the essential needs of the world's poor. They should be accorded a priority while the second pre-concept relates to the 'idea of limitation imposed by the state of technology and social organization on the environment's ability to meet present and future needs'. In fact, it is the limitation of the state of technology and social organizations which provides sustainability an inter-temporal relationship between human needs and human productive capacities. In other words, sustainability is being used as a sort of relationship between human welfare at different stages of human development, while the environment is being treated as a passive element in the entire process because needs are human-determined and limitations are also seen as human limitations.

This conception of sustainability given by Brundtland definition is social scientific in approach as it is popular among social scientists such as demographers and economists. It also focuses empirical attention on human demands and on the characteristics of technical and social innovations. Because of the controversial issue of natural resources being finite rather than infinite, this definition of sustainability in the social scientific terms is based on a very strong principle of inter-substitutability of resources. In fact, the principle of infinite inter-substitutability because of development of alternate means of recycling and disposal possibilities, provides a balance of human welfare across time (Norton, 2003).

It is further observed that the environmentalists may find it difficult to define sustainability simply as a matter of human technology and welfare. In the event, if sustainability is to be conceived as a simple problem of balancing welfare across generations, then human welfare has to be understood incrementally and interchangeably as viewed in the mainstream economics. In the incremental system, where all values are inter-changeable and all resources have requisite capital with adequate substitutes, environmental constraints may not be given special pre-empted significance. In short, an inclusive conception of sustainability should be able to judge the ability of the future to deal with pollution and waste and the declining stocks of resources. As some of the present activities are indicative of the risk of future shortages of resources and things for waste products, there can be a spectrum of risk severities such as extinction, biological impoverishment, cultural and political impoverishment, severe economic dislocation from resource scarcity and varying degree of higher real resource cause. This spectrum of risks as an incremental model appears irreversible in terms of explaining the process of species extinction, but it can be made comparatively reversible through having economic impacts of higher real resource cause and by having varying degrees of severity and reversibility. With regard to the decisions taken on reversible changes in the environmental policy, it may be good to apply the safe minimum standards of conservation as well as to protect the resources, natural process, species, etc., provided the costs are bearable (ibid.)

In the wake of the above, we as human beings are under obligation to protect the health and integrity of ecological systems that are based on the premise that the natural systems are self-organized. Towards achieving these goals, it is suggested to follow scientific contextualism for sustainability of the ecosystems as it is considered that for sustainable management, we have to envisage hierarchical-organized systems having human activities which impact them not individually but in larger parts. It is observed that technology and population growth have provided human cultures an ability to alter larger and normally slowly changing systems of ecology, climate and atmosphere as well as to initiate fluctuations in these systems (Leopold, 1939).

As such, in the contextual analysis, it is pointed out that individual behaviour is not the main focus of environmental ethics but the trends in behaviour are going to determine whether or not they will have inter-generational impacts. For example, if a farmer clears his wood lot to plant wheat, this may not be wrong as long as his land is on a highly roadable slope and in case his action is not followed by all of his neighbours. In fact, it is not the content but the context of his action that needs to be assigned a moral dimension.

Contextualism emphasizes on the self-organization of large systems as they can assimilate human impact, which is large but not infinite. Self-organized systems are autonomous in nature and sustainability in these systems can be conceived in terms of a relationship between dynamic human economic systems and larger dynamic but normally slower changing ecological systems (such as continuity of human life indefinitely, flourishing of human individuals and development of human culture) in which effects of human activities remain within limits so as not to disturb or destroy the health and integrity of the environmental context of human activities.

This proposition may define ecology 'as the science of the principles of the self-organization of nature'. As such, ecology and not economics become the focal point of the sciences because economic activities are to be conceived as a type of ecological activity that may take place within the economic system of one species to build up an ecological paradigm of environmental management. In this context, fivefold axioms have been suggested. First, it is believed that nature is more profoundly a set of processes than a collection of objects which are in flux. Secondly, all processes are related to all other processes. Thirdly, processes are not related equally but they unfold themselves in systems within systems, which differ mainly with regard to temporal and spatial states. Fourthly, the autonomous processes of nature are creative and represent the basis for all biologically-based productivity and finally, ecological systems which represent the context of all human activities are varied to the extent to which they can observe and assimilate human-caused deceptions in their autonomous processes (Norton, 2003).

In the above context, it is observed that a system is healthy if it maintains its complexity and autonomy/self-organization. In fact, self-organization is to be considered an essential part of health and integrity of an ecosystem. Thus, it is necessary for us to protect biodiversity for our future generations as a part of our obligations to them and even more important to protect the complexity of self-organizing systems involving complex processes of ecosystems.

It is argued that environmental policy should get influenced by both ecological and economic dimensions. Economic concerns are in the forefront when risks are not catastrophic or irreversible and when the areas affected are relatively small. However, non-negotiable inter-generational obligations acquire dominance when decisions are to be taken involving the risk of irreversible or catastrophic change in large-scale systems on which the human species depend (ibid.).

Notwithstanding the above observation for sustainable development of the different ecological systems for human welfare and for managing the health and integrity of the ecosystem affecting the larger systems of environment, one cannot ignore the damaging effects of environmental deterioration caused by large-scale pollution and wastage discharge for health risks of the individuals who are productive actors in the development processes, especially in the developing nations. This, because largely, the nation-states of these countries have to rely more on labour intensive rather than technological intensive systems of production. As such, the health risks of individuals not only need to be protected by developing a sound curative health care system, but also through strengthening of preventive medicine measures for sustainable development of these nations. In addition to the individual health behaviour risks, we shall also have to evolve comprehensive health care at the family and community level for strengthening the health status of populations inhabiting each nation-state for improving their quality of life.

In the subsequent analysis of environment and health, we shall make an attempt to discuss the debatable issue of the role of individual and society in maintaining and restoring the health of people vulnerable to various kinds of health risks that may lower productivity and affect the development process adversely because of environmental pollution and wastage disposal problems.

References

Brundtland Commission Report (World Commission on Environment and Development) (1987), *Our Common Future*, Oxford: Oxford University Press.

Chandna, R.C. (1998), *Environmental Awareness*, New Delhi: Kalyani Publishers.

Ekins, P. (1996), 'The Relationship between Economic Growth, Human Welfare and Environmental Sustainability' (Unpublished Ph.D thesis), London, Department of Economics, Birkbeck College, University of London, as quoted in Michael Redclift's 'Environmental Social Theory for a Globalizing World Economy', in *Environment and Global Modernity*, Gert Spaargaren, Arthur P.J. Mol and Frederick H. Buttel (Eds), London: Sage Publications, 2000.

Ekins, Paul (1991), "A Strategy for Global Environmental, Development", *Development*, Journal of the Society for International Development, No. 2, Rome.

Escobar, A. (1996), 'Constructing Nature: Elements for a Poststructuralist Political Ecology', in *Liberation Ecologies: Environment, Development and Social Movements*, R. Peet and M. Watts (Eds), London: Routledge.

Giddens, Anthony (1994), *Beyond Left and Right*, Stanford, CA: Stanford University Press.

Leopold, A. (1939), 'A Biotic View of Land', *Journal of Forestry*, Vol. 37, pp. 727–730.

Miller, Alan (1996), 'Economics and the Environment', in *Key Concepts in Critical Theory: Ecology*, Carolyn Merchant (Ed.), Jaipur and New Delhi Rawat Publications.

Norton, B.G. (2003), *Searching for Sustainability: Interdisciplinary Essays in the Philosophy of Conservation Biology* (Cambridge Studies in Philosophy and Biology), Cambridge: Cambridge University Press.

Norton, B.G. and Michael A. Toman (1997), 'Sustainability: Ecological and Economic Perspectives', *Land Economics*, Vol. 73, No. 4, Reproduced in B.G. Norton (2003), *Searching for Sustainability: Interdisciplinary Essays in the Philosophy of Conservation Biology* (Cambridge Studies in Philosophy and Biology), Cambridge: Cambridge University Press, pp. 225–248.

Redclift, Michael (2000), 'Environmental Social Theory for a Globalizing Economy', in *Environmental and Global Modernity*, Gert Spaargaren Arthur P.J. Mol and Frederick H. Buttel (Eds), London: Sage Publications, pp. 151–162.

Reich, Michael, Richard Edwards and Thomas Weisskopf (Eds) (1972), *The Capitalist System*, Englewood Cliffs, NJ: Prentice Hall.

Stone, C. (1987), *Earth and Other Ethics*, New York: Harper & Row.

3

Health Risks and Vulnerability of Humans to Environmental Degradation

Human survival to a large extent depends on security, health awareness and services and environmental quality. These are highly interrelated and need to be examined in the context of establishing a relationship between environment and health.

The health and integrity of an ecosystem, be it a human habitat or settlement, animal habitat or aquatic life (marine life), is for sustainable development of these species. The individuals inhabiting each of these ecosystems also get affected adversely in sustaining themselves as well as perpetuating their life as a consequence of the ecological fragility caused by environmental degradation, mainly due to overexploitation of natural resources for the progress of individuals, groups, communities and nation-states.

The vulnerability of ecosystems to environmental damage through globalized economic penetration and modernization processes has put the majority of the population inhabiting the planet to a health risk. It is in this context that the World Commission on Environment and Development in 1987 emphasized on having a more healthy environment which should focus on sound economic planning and maximizing the resource creation for improving the health and well-being of the people. However, it must be emphasized that improving the health of the people has to be the primary goal of any environmental programme. It is also observed that the preservation of the natural resource base may not result in preventive health or a reduction in morbidity and mortality. No doubt, the processes of modernization and industrial development may co-exist with environmental programmes directed towards resource regeneration, yet evidence suggests that in most parts of the world, industrialization has been observed to be harmful to people's health, particularly to the health of the poor, and the nation-states need to take special measures towards that.

In the developed Western industrialized nations, it has been observed that these have been subjected to risk factors emanating from over-industrialization and intensified urbanization processes. These have been referred as risk societies by Beck and Giddens while explaining the processes of post-industrial or later modernity period. Due to increasing uncertainties and complexities involved in development, the globalized economy through free flow of capital, technology transfer and availability of cheap labour has also penetrated into the developing nations from the developed ones. These are also now emerging as risk societies, causing environmental damage due to overexploitation of natural resources. These development processes have led to many health problems as a fallout of environmental degradation, wastage and air and water pollution. We do observe a high incidence of both communicable and non-communicable diseases in these nations. These can be increasingly linked to changes in the food habits, lifestyles, exposure to toxic wastes, and air and water pollution causing health hazards of various types.

Before we take up the issues and problems related to environment and health, which need to be analysed in developing nations like ours in different settings such as rural, urban, agricultural and industrial workplaces, mining, hilly regions, desert regions and flood-affected areas, etc., it may be apt to understand the concept of health, disease and illness to examine the perspectives on environment and health in a meaningful manner.

Concept of Health, Disease and Illness

Health is an illusive concept as various attempts to define it have remained unsatisfactory satisfactory. However, in a holistic sense, it must take into account the curative, preventive, promotive and rehabilitative aspects. That is why it is necessary to analyse the socio-cultural, economic, psychological and political aspects of health and disease, in addition to understanding their medical dimensions. Health has also been viewed as a situational concept as one may be healthy 'here' and 'now' and may not remain so 'there' and 'then'. This implies that health is to be conceived in a time framework and in the context of a socio-cultural or physical environment.

Over the years, certain indicators, however, have been evolved which have been applied universally in the case of healthy and non-healthy persons. To a lay person, health would mean a sound physical body that may help a person to perform day-to-day activities to the expectations of others. As such, the individual and others constituting a group become two important dimensions in conceptualizing health. Animals have a natural instinct to keep themselves healthy. Likewise, human beings also possess these instincts but they are mostly insensitive to them till they find an impairment in their daily normal routine functioning in relation to either self or others with whom they interact.

It is this individual versus group or societal context which needs to be examined to understand the concept of health from different perspectives.

It is difficult to conceive health without its contrasting concept, that is, disease and illness. In that sense, health should mean absence of disease or illness and logically illness would mean absence of health. In other words, health is to be conceived in the normal behaviour of an individual whereas disease or illness may reflect an abnormal state of an individual. However, it has been argued that due to the unobtrusive nature of normality, it is much easier to identify abnormal or illness behaviour. In this context, health has to be viewed as a feature of daily living and a person's display of his or her essential normality. Healthy people are normal and normal people are healthy (Dingwall, 1976).

However, we do observe that there are some common health denominators shared by members of a group or community. These are generally inclusive of high physical level of activity, a well-flushed body and an absence of pain. Further we need to understand the health standard of physical activity according to age and gender categories. There is also a need in a community or societal setting for an individual to have access to knowledge of what is normal or healthy as shared by other members or the collectivities. Besides, he or she should have adequate sources to realize that knowledge in action. That would mean in the context of developing nations, there is a greater need to utilize health knowledge and provide services to all strata of groups.

Health also needs to be differentiated from illness behaviour. In fact, health behaviour is a kind of activity undertaken by a person who believes in himself or herself to be healthy in order to avoid or prevent a disease, whereas illness behaviour is a sort of activity by a person who may feel ill for the purpose of defining that illness and seeking relief from it (Kasl and Cobb, 1966).

Though the concepts of health and illness may be distinguished from one and other in qualitative terms, in common usage, one may visualize them with a similar reference. That is why, according to WHO, health is to be conceived as a 'state of complete physical, social and mental well-being and not merely the absence of disease or infirmity'. It is observed that the human body's interaction with mind is a kind of separate entity which gives rise to emotions, sentiments and feelings. The adjustment of these is crucial for the health of the individual. On the other hand, disease has to be considered as a medical concept of pathological abnormality revealed by bacteriological analysis, whereby certain kinds of bacteria or viruses cause impairment of the human body and make it dysfunctional. Illness mainly refers to a person's subjective experience of absence or lack of health due to feeling of pain, discomfort, etc.

To say that a person is ill would mean not only the biological and physical consequences of organic malfunctioning, but a state that affects his or her social life in important ways. It is in this context that the determinants of health and illness have to be conceived in terms of social, economic, psychological and political dimensions. It has been reported that physical, cultural and social environment affects the state of health of an individual, family, group and society and these constantly influence a person's psycho-physiological energy system (Freeman, 1970).

As observed by Freudenberg (1978), the environment itself will be either health-promoting or health-negating because an individual brought up in a conducive environment may become an active contributor to the society, while one brought up in an unhealthy environment will prove to be a liability. It is also observed that there is a kind of relationship which exists between social environment and the cause and course of diseases. In this respect, social epidemiologists basing their analysis on interaction among the host (individual, families, groups, communities), agents (bacteria, viruses, germs, parasites, etc.) and environment (physical, social and cultural) have suggested an ecological perspective on health and illness. It has been suggested by Brown and Margo (1978) that individual behaviour should work hand in glove with the physical environment and social conditions of many important diseases. Social epidemiological studies by Penfold (1981) have suggested four important factors such as economic inequalities, gender inequalities, occupational risks and environmental risks in analysing health and disease patterns. In view of that, Allison (1992), in addition to the behavioural approach has suggested a structural approach, based on environmental and social factors related to health and illness for a balanced treatment.

As such, the environmental interventions become crucial and we need to alter those aspects of the physical and social environment that make the adoption of healthy individual habits difficult. Brown et al. (1974) identified some simple aspects of an individual's social environmental crisis such as bereavement, social isolation, moving out, etc., and suggested that these could have far-reaching influences on the health as without them, an individual would have remained well for several years. As such, there is a need to explain to the people through health education, the behaviour and conditions that should be beneficial to their health, but also help them remove health-damaging conditions by replacing them with the health-promoting ones. It is in this regard that the health professionals apart from focusing on preventive health measures should also consider a politico-economic focus (Mckinley, 1975).

Individual vis-à-vis Group or Societal-Oriented Perspectives on Health

Both the individuals and the societal or group factors have been employed as units of analysis in understanding the human behaviour in micro and macro sociology. Peter Berger (1963) provided an interesting sociological perspective on Man in Society (Macro Sociological) and Society in Man (Micro Sociological). While the former considers society as an external reality that may exert influence and coercion upon the individual, the latter focuses on internalization of the social world; that is, the structures of society become the structures of our own consciousness. In fact, the individual and the society seem to be two entities confronting each other. However, Durkheim provided a socio-centric perspective which argues that an individual does not have an entity of his or her own; rather, he or she is a product of the social world. According to Durkheim, society confronts us as an objective reality. In other words, our location or position in the specific sectors of the social system may pre-determine and pre-define everything for us from language to mannerism, from religious beliefs to our chances of even committing suicide and we have to conform to the prescriptions of the society. Even Levi-strauss has emphasized on a communication model to transform the external environment by an individual through exchange reward and communicative behaviour.

An individual, though, has his physical body but he cannot be conceived as an entity without understanding his or her personality structure. This is in part determined by bio-mental processes in interaction with an array of external stimuli existing in the vast field of social environment. The mediation of external organism between the stimulus and response promotes selective memory storage, perception and action on the part of individuals interacting with one another. These interactionist processes have been observed to operate even in highly rigid structures. Further, interpretive sociology also allows an individual to choose out of available alternatives for action as a rational human being. These theoretical formulations have implications on an individual's healthy habits, healthful living and to act in accordance with a situation while providing coping responses on the basis of his or her understanding, sharing of experiences, medical knowledge and cultural imperatives. Even Gandhian philosophical treatise suggests that an individual will have to bring about self-transformation in order to realize healthful living, adopt healthy practices and avoid health hazards. This kind of behaviour pattern by an individual to adopt healthy practices and avoid morbid practices may be achieved through various techniques and methods. For example, the psycho-analytical framework within the Freudian psychology focuses on the response structure of the individual patient, suffering from stress or anxiety, by providing control to stimuli given to the patient in order to bring about normal functioning potentialities of the individual. This approach makes no attempt to control deviance by manipulating the social structure. The symbolic interactionist perspective also views man as a rational human being with a capacity to structure his environment and relations with others. But, unlike psychotherapeutic techniques, it does not provide any treatment. However, interpersonal therapy and transactional analysis are being used now to affect behaviour change learning theory in social psychology, which focuses on conditioning and desensitization to help patients acquire new repertoires in relation to certain specified social and environmental stimuli. Those patients who make a better adjustment to the environment are rewarded, while patients who are exposed to the painful stimuli slowly may find these threatening but these may help them to overcome their fears. However, the learning theory has been observed to be more effective in voluntary situations.

In view of some of these above perspectives, scholars have identified individual-oriented factors such as perceived stress (Mechanic, 1978) and anxiety (Kosa and Robertson, 1969) as significant for causing illness. On the other hand, a number of other factors such as inability of the sufferer to accept his or her normal symptomatology (Zola, 1965), social group structure and individual medical orientation and cultural gap between the low status users and professional medical practitioners to avail of medical facilities by the poor and disadvantaged groups have been identified with regard to the pattern of seeking medical care from the available resources, in the event of sickness or illness (Rainwater, 1963; Strauss, 1969).

Though primitive human beings relied more on their instincts to stay healthy as they could not comprehend the functioning of the human body, magic became an integral part of the beliefs about the causes and cures of health disorders (Dubos, 1969). Such a belief is still being held by a large population, especially in the developing nations. However, during the medieval times, illness was conceived as a sort of punishment for one's sins and the care of the sick was considered as some kind of religious charity, but in the contemporary context, illness is regarded as a state or condition of suffering as a result of a disease or sickness.

There emerged several perspectives in sociological literature on health, disease and illness. We shall highlight some of the dominant perspectives on Sociology of Health which have been built around individual or group-oriented factors on health.

Parsons (1951) was the first among the sociologists to develop the concept of the sick role. His functional perspective had a lot of influence for long on the sociological research related to the health field. Marx, on the other hand, emphasized on the productivity of man, but did not specify much on the conception of health or sickness. However, the Marxian perspective, of late, is being implied in the political economy of the health approach especially related to poverty groups.

Parsons (1951) viewed illness as a deviant state brought about by disruption of normal behaviour through disease. He conceived illness as a deviant behaviour within the sociological definition of deviance as a sort of any behaviour violating the social norms within a given social system. This functionalist approach to deviance led to his conception of the 'sick role' concept, which considered sickness as a disturbance in the normal condition of the human being both biologically and socially. His main postulation is that sickness does not imply experiencing the physical condition of a sick state. Instead, it constitutes a social role as it involves behaviour based on some kind of institutional expectations, reinforced by the norms of society related to these expectations. A major societal or institutional expectation is that a sick person is not in a position to take care of himself or herself, as such he or she needs to seek medical help and cooperate with the medical specialists to get well. More specifically speaking, Parsons' 'sick role' may be conceived as having four main components:

1. The sick person is exempted from normal social roles depending upon the severity and nature of illness.
2. The sick person is not responsible for his or her condition.
3. The sick person should try to get well since being sick is undesirable.
4. The sick person should seek technically competent health care and co-operate with the physician.

The sick role concept has been observed to be a useful sociological approach as it takes into account the social roles, attitudes and activity of both the patient and the doctor in a given situation. But, it was found to be clinically-oriented and failed to explain the variation within illness behaviour, applicable to chronic illness. Besides, it is not in a position to explain the same in a variety of settings and situations affecting the doctor-patient relationship and the behaviour of lower class patients (Cockerham, 1978).

It has been observed that in the context of developing countries such as ours, majority of the poor people cannot afford to seek medical help to get well or they do not have a necessary will to go to the physician. Instead, they expect that medical care should be brought to their doorsteps. Despite this limitation, this concept has been observed to be useful in the ideal type situation. In his subsequent observation, Parsons was aware of the limitation of his concept to include the whole range of phenomena associated with the sick role.

Another significant perspective to explain illness behaviour is based on Becker's 'Labelling Theory' (1963), which emphasized that deviant behaviour by one person or social group might not be conceived as such by another person or social groups. In other words, two persons having similar symptoms may behave differently. Freidson (1970) and Lipovski (1970) mentioned that illness as a deviant behaviour was relative and needs to be conceived from the perspective of labelling theory. This has significance for seeking medical care behaviour but is inadequate to provide a complete understanding of the sick roles or illness behaviour. The underlying assumption in this theory is that all behaviour is self-directed on the basis of symbolic meanings that are shared, communicated and interpreted by human beings by interacting in different social settings. Being thinking organisms, they have a tendency to choose their behaviour instead of reacting mechanically to the effects of social processes.

Freidson (1970) brings in the concept of legitimacy to make a distinction among sick roles of various kinds. In his analysis of legitimacy, he explains the differential behaviour of deviance suffering from minor or serious diseases. For example, conditional legitimacy allows a deviant a temporary exemption from normal obligations and provides privileges to him to seek medical aid in order to get rid of the deviance, whereas the unconditional legitimacy exempts a deviant permanently from normal obligations of his role and grants additional privileges to him to manage the hopeless nature of the deviance. This perspective has been considered by some as better than the concept of the sick role of Parsons in explaining different types of illness and in emphasizing that illness is a socially-created label. However, Friedson fails to explain the differences in the way people defined themselves as sick and needing professional medical care. Further, this analysis of illness behaviour restricts oneself to the framework of a social event or setting and excludes the biological properties specified by the physicians to identify the etiology of disease for a complete analysis of illness behaviour.

In the context of the developing nations, another perspective that became significant in explaining illness or disease behaviour is that of ethno-medicine. This became relevant and valid in the context of other kinds of knowledge like folk medical knowledge than Western scientific knowledge. It was observed that though Freidson (1970) following the labelling theorists, recognized the relevance of cultural pluralism, his cultural groups remained subservient of their values. Because of these limitations, the ethno-medicine approach advocated by Dingwall (1976) appears more relevant than the Western scientific approach as it provides a wider perspective to the understanding of concepts of health and illness and organization of health services by including within its purview the folk and indigenous medicines, and even self-medication practices of the lay people.

According to Dingwall (1976), illness can be conceived as a form of failure at everyday life, a disruption in the 'familiar and taken for granted' state of affairs between the subjective experience of one's own knowledge about the normal experience or the conduct determined by some competent members of a collectivity. In other words, ethno-medicine considers medical knowledge invoked by the physician to interpret a patient's behaviour similar to the knowledge invoked by a lay person about any other person's behaviour or conduct. These are based on the knowledge of members of some collectivity who draw on it to make sense in their social and natural world and on the content and organization of that knowledge (ibid.).

Another perspective that acquired significance in the field of preventive medicine is given by Rosenstock (1966). This 'Health Belief Model' is based on social psychological factors which have been identified on the basis of an individual's perception that he or she is susceptible and the occurrence of the disease may have severe implication of a personal nature. This perception is likely to indicate him or her to accept or adopt preventive practices to avoid illness. This perspective is applied mostly to prevent situations involving voluntary behaviour such as smoking, drug addiction or alcoholism, etc. But it is observed that most of the people seek medical help or health care services mainly when they have clear and definite symptoms of the disease. As such, one has to see whether people would be motivated to avail medical care services in order to improve their health and avoid future disease or not. In this regard, Mechanic (1978) observes that the main difficulty in preventive medicine is based on the fact that commonsense approaches do not always match the clinical approach. However, commonsense often determines whether one should seek health services or not. But this model has some merit in the sense that even when an individual is perceiving personal susceptibility, he or she may not take action unless it is also felt that being ill will result in serious circumstances. This subjective assessment of the health situation may thus become a critical variable in the utilization of health services by an individual in a given society.

Some of these significant perspectives on health have been in practice in health sociology. However, there are many variations of these, which may be conceived as derivatives of the main theoretical frameworks outlined above. These perspectives have been observed to influence the individuals and groups located in different physical and cultural settings. Some of the aspects of these perspectives will be useful in analysing various aspects of environment and health such as personal hygiene of an individual, insanitary living conditions of people, insanitary practices of individuals and families, air and water-borne diseases causing damage to the environment due to pollution of air and water, chemical substances and radioactive nuclear radiations, damage caused by destruction of fauna and flora, etc. An attempt will be made to highlight some of these issues in the next section.

Environment and Health: Visible and Invisible Health Risks in Different Socio-cultural Groups

In the Western industrialized nations, many of the health problems have been observed to be the outcome of environmental hazards. For example, the high incidence of heart disease and cancer has been largely linked to diet, lifestyle, exposure to toxic wastes, etc., which perhaps could have been controlled by suitable changes in a given environment. But in a country like ours, people are prone to illness and disease due to conditions associated with poverty, malnourishment, overcrowding, congestion, etc., especially in the urban slums. In their case, perhaps it is difficult to change their environment.

In his interesting paper, Jill-Carr Harris (1992), while analysing environment and health, mentioned about two perspectives. One is the environmental perspective on health, while the other is referred to as the health perspective on environmental hazards. According to the first perspective, the environmental programmes in our setting have not been given adequate attention to health and other social costs. Environment is treated as an external physical entity having only a functional or economic bearing on people, but it is to be observed that many of the infrastructural projects such as constructing dams, etc., may lead to increased morbidity or mortality rather than reducing them. We cannot treat the Western type of environmental measure by introducing high technology when we have availability of huge raw materials and cheap labour. Such an approach may have a negative effect on employment and overall productivity. Apart from these deleterious effects of industrialization on the people, there is a tendency to consider environment having a set of finite natural resources. This type of bias is mainly because of adopting the reductionist approach to environmental sciences. In fact, environment should be conceived as an integrated, physical and social setting in which people live and the sociology of environment must focus on the effects of environment in changing health status by analysing the impact of degenerating ecosystems on people's health.

The other perspective (health perspective on environmental hazards) is due to the promotion of community health in the Third World nations during 1970s because of the availability of large literature on the social and cultural aspects of a de-medicalized health system. It is observed by Harris (1992) that generally health sociologists tried to view environment similar to the physical condition of poverty or as a set of practices relating to hygiene and sanitation that perhaps can be easily managed. However, the poor people are constrained to change the environment to manage their health. We cannot reduce environment to the physical conditions of a particular disease as was witnessed in the case of malaria, which resurfaced in 1977 even after two decades of DDT spray.

There is a need to view the environment as an ecosystem or habitat and an attempt in this direction was made in the 1980s by the vector controlled research centres in Puducherry through bio-environmental control. Even in the context of the health perspective on environmental hazards, we cannot equate environment with a set of unhygienic practices. This has been observed in the case of programmes on water and sanitation. With regard to cleanliness, purification and hygiene, one has to take into account the traditional beliefs and customs of the Indian society to place them in a proper socio-cultural context. It is in this sense that the environment has not to be treated simply as a question of hygiene, but as a part of the integrated system of physical and social relations. As such, we cannot always assume that ill-health is manifested by disease. In fact, ill-health should be conceived usually as a condition of weakness or stress and without attaching any importance to the symptoms. This should also be considered as some sort of loss of balance or homeostasis due to change in the environment. In fact, we need to understand the person's changing environment in order to reduce the risk of health hazards.

It is, therefore, necessary to treat the environment in which we live as an ecosystem or as integration of physical and social relations, which implies that the conditions of ill-health are due to multiple factors or events. Thus, it may be difficult to establish the correlation between environment and health. In this context, we may have to shift from the clinical testing of the disease and its causal agents to the deconstruction of some of the social and physical relations to focus also on preventive health measures (Harris, 1992).

Environmental damage or hazards have placed a large number of people at invisible and visible health risks. This has been pronounced mainly due to increasing modernization of agriculture and industry as well as due to overcongestion and crowding of people and increasing urbanization. For example, a large number of people in the workplace, rural areas and urban areas are susceptible to invisible health risks due to a stressful life such as hypertension, diabetes, passive inhaling of smoke causing respiratory diseases, heart stroke, anaemia, depression, cancer, etc. (hypertension and diabetes have been considered as silent killers), whereas there is manifestation of visible health risks in the occurrence of parasitic and vector-borne diseases such as malaria, filarial, tuberculosis, small pox, diarrhoea and dysentery, cholera and protein-deficiency diseases.

India being largely an agricultural and biomass society, nearly 70 per cent of our population lives in villages and are engaged in agricultural and allied agricultural operations such as dairy farming, fishery, poultry, etc. As such, they are subjected to exposure of various risk factors both in their home and work environment, whereas those in the urban areas are mostly engaged in occupations related to production, trade and services of various kinds. Thus, they are also exposed to various risks in their home settings. In the case of majority of slum-dwellers, they live in highly congested places and are without any water and sanitary facilities. As such, they are susceptible not only to health risks in their home settings but also have to confront many occupational hazards causing ill-health and disease in their workplaces.

In view of the above, it may be apt to consider 'environment and health' in the context of home, rural and urban environments and to identity social and cultural factors having a bearing on these.

Home Environment and Health

In the home setting, the most important aspect is that of personal hygiene related to maintenance of cleanliness, bathing, washing of hands before and after meals, oral hygiene (dental care) and availability of potable water and adequate toilet facilities for maintenance of sanitary living conditions, including disposal of garbage and other chemical waste material, etc. These measures if taken adequately by individuals and families can prevent many diseases or ailments. As such, it is essential to have a good physical surrounding such as a good house, proper drainage and sewerage system so as not to allow the water stagnation for breeding of insects and parasites, proper ventilation of the house to have fresh air and sunlight into the rooms. The other significant factor to be kept in mind is to have availability of safe drinking water either through the pipeline supply after chemical treatment of water, or after the decontamination of supply of water through wells in the villages. Apart from these physical facilities, the other non-physical dimensions such as psycho-social are also important to keep the home environment healthy.

Healthy habits can be inculcated not only among adults but among the children by family socialization processes to help them acquire a set of habits and behaviour. It is well known that a good healthy home environment can have linkages with people in the community, outside the natural surroundings and the larger society for achieving good health. In this regard, women have to assume a significant role in disseminating health information to children through their proper mothering skills for providing a healthy environment. For example, it has been observed in Kerala that the high health status in that state is related to high literacy level of the women. Even, UNICEF (United Nations International Children and Education Fund) through its various welfare agencies promoted 'care for child' motto, to provide child with a productive life. In the context of rural dwellings, there is a dire need to separate living rooms from that of animal sheds. This has to be taken up on a priority basis to prevent environmental health hazards, in addition to supply of safe drinking water to the families living in villages. There is also a need of providing environmental sanitation by having rural latrine programmes as in most of the villages in India, people are still going to the fields for easing out and defecation. This has serious implications for contacting diseases through parasite germs. Women and children are the most sufferers of these kinds of cultural practices related to defecation. (In this regard, a significant report of Conference on Social and Cultural Factors in Environmental Sanitation in Rural India was brought out by the Publication Division of

Government of India in 1956. This was based on a major effort in the field of public health by having collaboration between the health administrators and social scientists in the research-cum-action project on Environmental Sanitation (Singhur in West Bengal, Najafgarh in Delhi and Field Study Area of Planning-cum-Action Research Centre, Lucknow). A number of social and cultural factors which were related to the set habits of villagers were identified and based on that the rural latrine programmes of (compost pits) were initiated in certain parts of the country.

Rural Environment and Health

Although many health risks, diseases and illnesses cut across the rural and urban environments, yet it would be appropriate to highlight some of these in the two settings as there are basic differences observed in the occurrence of health hazards in these environments.

In the rural environment, especially due to introduction of modernization processes such as industrialization and urbanization, infrastructural development such as construction of dams for irrigation purposes, power generation plants and road networks have affected people's relationship with land, technology and production. These processes have increasingly marginalized those involved in the subsistence level of economy, especially the landless, because their means of survival have been appropriated by the capital investment in human and natural resources. As a consequence of that, they suffer the highest rates of mortality and morbidity.

In rural areas, poor housing conditions indicate a poor living environment. Despite an increase in the number of cement houses, there has been an increase in the vector diseases such as malaria, etc., because most of these do not have a living space and their location is not having access to certain facilities. Most of the villages do not benefit from modernized agriculture because the capacity for local resource generation is low due to poor facilities and services. For example, the water resources are, by and large, unprotected and majority of the community wells are polluted. These wells provide a good location for mosquitoes to breed and cause malaria. Even water tanks and tubewells may encourage such breeding if the water remains stagnant in these places. As far as possible, housing sites should be kept away from stagnant water. (It has been suggested that the construction of settlements near reservoirs, canals, water channels and tanks for purposes of irrigating land should be kept away for more than 3 kilometres as the mosquito-carrying parasite cannot usually fly more than that from its breeding ground. But, if a settlement is closer than this to the irrigation source, it is susceptible to health hazards). Besides the irrigation source, another factor that is conducive to vector breeding is the cohabitation of humans and animals. This has been observed in most of the village settlements because of overcrowding in homes having an average area per person of about 5 square metres. Finally, the rooms are usually small and windowless and these are conducive for the transmission of parasites (Harris, 1992).

The latest source of health hazards among the rural households is non-availability of safe drinking water. Though in our ancient texts (*Sushruta Samhita*), various methods to purify water have been mentioned, they are found to be inadequate in water-related diseases such as amoebic and bacillary dysentery, cholera, diarrhoea, hepatitis, typhoid and guinea worm. The low birth weight or malnourished children have been observed to be at risk to catch infections because of their weak bodies.

Though attempts have been made by the government to provide safe drinking water to the rural population, not more than one-third have access to it. For nearly half the villages in India, people's only source of water is observed to be cholera endemic. We do not have sufficient drinking water available in India as has been observed in the sense that the Indian population consumes not more than 1 to 2 per cent of the total water supply. In fact, there is water scarcity which confronts us as a reality. Even the construction of dams is not boosting drinking water supply and the construction of tubewells is mainly for generation of water to be utilized for the irrigation of cash crops and hybrid species. This limits the availability of water sources and people are compelled to use the limited water supply for their several needs such as feeding livestock, washing, drinking, irrigation and watering the kitchen garden, etc. As a result of this, personal hygiene gets neglected. People are subjected to health risks by diseases such as scabies, trachoma, leprosy and conjunctivitis. These health risks are pronounced in lower caste households as they are generally denied access to safe water sources due to the dominance of higher castes.

Another problem in the villages is that of waste disposal. This has come to the forefront due to increasing population and diminishing land area. There is increase in the excreta-related diseases such as hook worm, etc., which affects the barefoot workers most. It is reported that hookworm is mainly responsible for a high proportion of rural women suffering from anaemia. Even after efforts made by the government to have rural latrine programmes, it is reported (Banerji, 1982, as quoted by Harris, 1992) that still 8,790 villages have no latrines at all and those who had bucket latrines did not improve environmental hygiene. In the recent past, common toilets promoted by Sulabh Shauchalaya and pit latrines using minimum water have been introduced to reduce water and excreta-related diseases. The sanitation conditions of these have also been observed to be worse. It is reported that only 4 per cent of the rural population and 38 per cent of the urban population have access to sanitation facilities such as private latrines and closed sewerage system (Sheth, 1997).

It is also reported that due to the involvement of women in rural areas to collect firewood and fodder, they are forced to change the food habits of their families as they are compelled to ignore their own nutritional intake to provide more for their other family members, especially children. Consequently, they become anaemic. They are also involved as agricultural labourers apart from doing the job of biomass collection. As such, they have hardly any time to rest and it is acute during the menstrual cycle and pregnancy. This is a serious matter of concern and is responsible for a large percentage of maternal mortality among anaemic women. Besides, it increases the risk of death during childbirth. There is even a problem of leucorrhoea or white discharge observed among pregnant women in their second trimester, in girls at puberty and in lactating mothers. In addition, women working in the construction activities have been observed to suffer from anaemia, malnutrition, calcium deficiency and are subjected to abortions being performed with unclean apparatuses.

These gynaecological problems need to be analysed with environmental factors in and around the home. Most of these problems are a result of lower status of women in the household, as that invariably denies them privacy in functions such as bathing and toileting, especially during the menstrual cycle. Such genital infections often take place due to lack of clean water, sanitary textile (napkins) and insanitary delivery being conducted in cowsheds or due to the transmission of sexual diseases. It is a stress-free environment that is needed to reduce the incidence of pregnancy and child-bearing disorders (Harris, 1992).

Alcoholic addiction is becoming increasingly an environmental factor, as observed especially in the rural areas. Those addicted become chronic in due course of time. Their condition of malnutrition further subjects them to mal-absorption of major nutrients, increases the levels of violence within the household, aggravates the condition of poverty and leads to a higher incidence of labour disorders. Women, especially in the tribal belt, are also addicted to alcoholic intake and this has an adverse effect on their menstrual cycle. Children of such parents (both men and women) who are addicted to high intake of alcohol have been observed to suffer from congenital heart ailments, poor coordination, hyper activity and other emotional and mental problems.

In the work environment of people in the rural areas, especially farmers, agricultural labourers, cottage and small scale producers and especially those involved in natural resource industry, there is a need to look for health, sanitation and other amenities of welfare in their workplaces as they are often subjected to environmental devastation caused by the production processes leading to destruction of land, forest, mineral wealth and water. For example, agricultural labourers bear a risk of poisoning while having direct contact with pesticides and fertilizers. Women labourers are more vulnerable to the effects of pesticides and other toxic agents, especially during pregnancy and after childbirth. It is also observed that workers in processing units such as quartz grinding and *beedi* rolling, etc., are likely to have a high incidence of tuberculosis and silicosis. Even mine workers, especially those involved in coal extraction, are not only subjected to accidents but are more likely to suffer from chronic poisoning and early mortality. It is observed that pesticide contamination in our setting is alarming and has an effect on microbiota, ground water, fish, animals and humans because a large number of rural workers use these toxic substances in their crafts.

Besides, the farming environment has been polluted considerably as the Green Revolution inputs such as hybrid seeds, chemical fertilizers, pesticides and artificial irrigation have given farmers a higher return of agricultural produce but at the cost of long-term productivity of land. This has resulted in soil erosion. There is a need to rely more on genetic diversity by use of animal and farm yard manure, integration of trees and animals, birds, insects and aquatic life into cropping patterns to protect both the soil and plants effectively for accumulation and cycling of natural nutrients. In the agricultural operations, the use of mechanical equipment such as threshers and shaff cutters not only leads to labourers being injured on the job, but in many cases, they are incapacitated. Even workers in the mining industry who are involved in stone breaking are likely to suffer from eye ailments, respiratory problems, injury to their hands and feet, etc., especially those involved in blasting of boulders, etc., with explosives.

In the hilly regions, women engaged in excessive labour to cultivate food crops, collecting biomass fodder and fuel are the worst-affected persons so far as their health is concerned. Besides, they are subjected to a high incidence of worm infestation from the contaminated water. A large number of them are highly susceptible to tuberculosis, while women also suffer from anaemia and gynecological diseases. In the desert or drought-prone areas, due to poverty and non-availability of food, the diets are deficient in calories, proteins and Vitamin A. It is reported that in such regions, protein energy malnutrition is observed to be common among young children while Vitamin A and B complex deficiencies are likely to occur in the older age groups. In the flood-affected regions, mainly because of large tracks of land under waterlogging for a significant part of the year, there is malnutrition observed among people due to non-availability of sufficient food. Even due to lack of potable drinking water, there is manifestation of gastroenteritis resulting in higher mortality among children. Further, due to increased humidity, there is occurrence of vector-borne diseases such as malaria especially when people are already undernourished due to the onslaught of floods.

Even due to infrastructural development such as construction of dams and rapid industrialization, mines, etc., there is an adverse environment not conducive to the health of all those who are at the subsistence level of economy. In fact, there is a change observed in the relationship between people and land, production and technology and as a consequence of that, there is an increase in the mortality and morbidity rates of people in the rural environment (ibid.).

Of the three major contributing factors to the emergence of health hazards in rural environment – water pollution, air pollution and intensive use of chemical fertilizers and pesticides – water pollution is reported to cause most of the diseases. WHO statement issue (July 2011) in its report 'Sick Water' points out that 90 per cent of waste water discharged daily in developing countries is untreated. Nearly 2.2 million people in a year die from diarrhoeal diseases caused by unsafe drinking water and poor hygiene. Nearly 1.8 million children less than five years of age also die every year from water-related diseases. In the context of the Indian population, it is estimated that nearly 37.7 million persons are affected by water-borne diseases annually, of which 1.5 million children die of diarrhoea alone. Until the Tenth Five-Year Plan, the state itself and even the private bodies failed to provide safe drinking water to nearly 700 million rural population in the country. The major contributors to the contamination of ground water are pesticides, sewage, nutrients such as phosphorus and nitrogen, synthetic organics, acidification such as sulphur dioxide from power plants, heavy industry like steel and motor vehicles, chemicals in drinking water, fluorides, arsenic, lead, petrochemicals, and chlorinated solvents such as metal and plastic effluents, etc. The major water-borne diseases caused are bacterial infections, typhoid, para-typhoid fever, cholera, dysentery including amoebic dysentery, jaundice or infectious hepatitis and polio myelitis, etc. On the other hand, the air pollution causing health hazards in the rural environment is mainly due to firewood burning for cooking purposes as well as causing health problems due to congestion of rooms and lack of ventilation in living rooms. Many people are victims of respiratory diseases as well as eye ailments in the villages.

Recently, the burning of rice husk and wheat stalk in the agricultural fields has also caused eye ailments such as trachoma, etc., and asthmatic conditions. A large number of truck drivers who hail from rural areas because of their multiple sexual contacts have been affected with HIV positive virus, subjecting them to the burden of AIDS (Acquired Immune Deficiency Syndrome). As carriers of this virus, they also pass it on to their wives and newly-born children (NACD estimate: 88.7% adults 15–49 years, 7.5% aged 50 years and above and 3.5% HIV children below 15 years). There is even an increase in the prevalence and incidence of tuberculosis cases in the rural areas of our country. It is reported that as of 2009, 17 per cent of TB patients are known to have HIV status (total rural-urban population) and 12 per cent of tested TB patients are HIV Positive.

In India, adult HIV prevalence has dropped from 0.45 per cent in 2002 to 0.29 per cent in 2008 to it lowest-ever number of 2.27 million (all states except Andhra Pradesh are reporting less than 1% infection, while eight states have HIV prevalence greater than the national average). Unprotected sex (87% heterosexual and 15% homosexual) remained the biggest source of infection and women accounted for nearly 39 per cent of HIV burden in the country.

Urban Environment and Health

After the gravity of health hazards and risks in the rural environment, we observe that with increasing urbanization of our country, which is nearly 30 per cent of the total population, we are facing degrading environmental problems in urban areas especially among the megacities, large populated cities, towns, etc. They are confronting decay due to polluting vehicles, traffic congestion leading to the release of carbon monoxide, unburnt hydrocarbons and suspension of particulate matter in the air, etc., causing many respiratory diseases.

A recent survey done by the Union Urban Development Ministry (as reported by Azad India Foundation in May 2010) on safe sanitation practices in 423 Indian cities, has brought out that on parameters such as absence of visible open defecation, proportion of total human excreta generation safely collected, proportion of treated waste water that is sampled, etc., Chandigarh tops the list while Churu in Rajasthan is at the bottom. Newly 190 cities mostly from Bihar and Uttar Pradesh are on the brink of public health and environmental decay. Delhi in spite of being the capital of the country is no better. It is reported that 90 per cent of Delhi, which is covered by Municipal Corporation of Delhi, has been put under the black category, requiring considerable improvement.

Besides that, it is estimated that each urban resident is likely to generate 350 to 1,000 grams of solid waste every day. Big cities like Delhi generate 4,000 tonnes of waste everyday, Ahmedabad 1,280 tonnes, Bangalore 850 tonnes, Mumbai 6,050 tonnes and Kolkata 3,150 tonnes of waste (Sheth, 1997).

As reported by Azad India Foundation (May 2010), during the last three years, the waste import to India increased by 48 per cent. In 2009, 6.4 million tonnes of hazardous waste came to India from the West, while 5.9 million tonnes was produced domestically. Much of this waste was in the form of metal, electronics and plastics, likely to be contaminated with lead, mercury and other toxics, causing serious illness and environmental damage. Most of this waste enters through ports of Mumbai, Chennai, Kolkata, Cochin and Vishakhapatnam and there is no radiation scanning technology available in some of these ports. These need to be monitored by the government to check the import of hazardous waste

Even the Municipal Corporations of big cities, despite deputing a large number of their employees in the task of cleaning and disposal of waste, have failed to completely dispose that of. Only two-thirds of the total waste is taken care of and most of the municipalities fail to keep roads and streets clean, collect garbage from public places and colonies, ensure its effective disposal and clean the sewerage. We have also not properly kept public latrines in most of the cities. Increasing pressure of population in most of the cities has resulted in weak infrastructure and these municipal bodies do not have adequate resources as well as technical knowhow to tackle these environmental problems. As unlike Western countries, we do not have the technical competence to recycle waste and generate energy from it. All these problems have caused environmental degradation of India's urban environment.

In addition to this gruesome situation, especially in the urban slums which are growing in large numbers due to migration from rural areas for better job opportunities in bigger cities (nearly one-third of population in big cities like Delhi, Kolkata, Chennai, Mumbai, etc., is reported to be that of slum dwellers), there is an acute shortage of safe drinking water as every city in every state has recklessly exploited the below surface water without any plan for its replenishment. The water scarcity is getting worse and worse due to the growing population as the demand for a consumerist lifestyle is on the rise, especially among the affluent sections, who need more bath tubs, showers in the bathrooms and toilet commodities in the houses.

Due to the fall of water table in most of the cities like Delhi and Kolkata over time, it has become a matter of serious concern for the future generation. More than that, the most telling effect is that of the degrading properties of drinking water causing various kinds of health hazards, especially diarrhoeal and dysentery problems. For example, in West Bengal districts, it was reported that drinking water contained more arsenic than the maximum limit prescribed by WHO. Even the Kolkata metropolis, according to a study, reported that the city is sitting on the 'brink of an arsenic bomb', despite the fact that its water resources are mainly dependent on tubewell water (ibid.).

As regards air pollution in the cities, the World Bank has estimated that the health hazards and environment costs due to vehicle pollution during 1991–92 were in the range of ₹ 1,810 crores and Rs 7,357 crores. Due to excess of air pollution, nearly 7,500 people are killed in New Delhi alone and there are additional 40 lakh admissions into hospitals in the country due to contaminates in the air. Also, 70 per cent of all air pollution is reported to be caused by vehicular sources.

Air pollution is also observed to be a key problem in slum areas due to indoor cooking with the use of dung or wood fuel and excessive invariants of dust at workplaces, especially for children who are exposed to the polluting industries in their proximity of living areas. As such, the workers in the small-scale industries where there are no health and safety measures also suffer from health hazards. In the case of the leather tanning process, the resultant effect of water pollution has been observed significantly. There is a need to provide pollution control regulations even for small-scale processing industries (Harris, 1992).

Unfortunately, even cities like New Delhi, Bangalore and Pune, which have vast spaces and greenery, are facing the onslaught of urbanization. Because of the vote bank politics, the political leadership through their municipal bodies has failed to provide clean cities and basic civic amenities. There is mushrooming of illegal or unauthorized constructions by builders which mar the city's landscape, affect its traffic and exert pressure on the urban system to cope with the growing demands of water, housing and toilets and disposal of massive garbage, increasing sewage and waste. Besides these problems, the municipal bodies have failed to check the growth of slums. This has further put a burden on the available infrastructure like roads, water and power, etc. Chemical industries, especially around cities like Ahmedabad, Baroda, Agra, Bhopal and pollutant industries around Delhi, Thane and Howrah, have further degraded such cities. Despite a number of NGOs which have come forward in these cities to tackle some of the environmental health hazards, much headway could not be made in this direction. It is observed that there is a need for a strong political will, huge resources, urban development on scientific lines and citizens participation in keeping the environment clean and urban resources free from misuse or overuse (Sheth, 1997).

Lack of sewage disposal and treatment of water make it difficult to provide clean drinking water, especially to the slum dwellers. Even the disposal of night soil is a major problem as not more than one-third of the population has one or the other kind of latrines for the disposal of excreta (21% use common flush toilets, 7.2% use flushes connected with sewer and 5.7% use septic tanks, as revealed by the 1971–72 National Sample Survey). In order to tackle some of these health hazards in the urban environment, both the state and the NGOs should play an important role in managing people's health and well-being in different physical and social environments (Harris, 1992).

It may be worth mentioning here that one of the NGOs in our setting, led by a trained sociologist Dr Bindeshwar Pathak, did an excellent social engineering of sanitation work not only in villages and towns but also in slum areas of mega cities and big urban towns by providing Sulabh Shauchalayas (latrines). His efforts are laudable and have been recognized by the national and state administrations to control and check excreta-related diseases. However, the main difficulty observed in the maintenance of these Shauchalayas is scarcity of water in urban slums and big cities.

It is also observed that the health and morbidity patterns in urban slums are even worse than those in the rural areas. For example, at Dharavi in Mumbai, which is the largest slum in the world, more than half a million people are living within a relatively small area. In greater Mumbai, there is the highest number of poor persons and it is estimated that by 2000, 7 out of every 10 persons may be living in subhuman conditions. (Slumification and urban decay are not only as a result of population growth and urban migration, but these are mainly due to skewed development which is in favour of the rich). Even in Delhi, where there is high expenditure on services, we have an acute shortage of clean drinking water and environmental hygiene for more than half of its population. The air pollution, because of increasing number of industries and thermal plants and untreated sewage, is very high.

Broadly, there are three main reasons which have impeded environmental health in urban slums. These are: (1) unequal access to land, public utilities and to urban infrastructure; (2) proximity to an industrial or other polluting source without the provision of medical facilities which can tackle chronic or acute toxicity, respiratory diseases and other industrial-related health hazards; and (3) insufficient living spaces for the population having an adverse impact on the social and political environment of the home (ibid.).

No doubt, many attempts have been made by the NGOs as well as by the government for slum renewal and upgradation since 1977, but there are many aspects of health and environment that have not been addressed adequately. (The government launched the urban basic services scheme under the Seventh Five-Year Plan. This was basically to integrate water supply and sanitation services with basic health care. The package of services included immunization of children, improved feeding practices, home-based diarrhoea management, drinking water supply, environmental sanitation and family welfare through community involvement). For example, the environmental health problems associated with pollution and chronic toxicity are not very well developed. The polluting industries, due to increase in the number of factories and small cottage industries, are destroying the fragile ecosystem by contaminating the ground water or if an industry is located in poor urban areas such as we had the Union Carbide in Bhopal, then the production process of such an industry may prove harmful to the local residents (ibid.).

There is an urgent need in such situations to educate the public and to provide appropriate safeguards. With rapid expansion of urbanization and industrialization, the emerging problem of health hazards is about chemical pollution affecting the water and environment. Even environmental health has to take care of the regulation of organic rather than chemical waste, as we have a high prevalence of water/excreta related diseases. In this context, both organic pathogens and chemical substances need to be taken care of in urban areas as most of the poor habitations are in close proximity to industries producing them.

We also cannot ignore the critical link between the health of the family and the environment. For that, we need to have female education and 'check child mortality'. Most of the migrants from rural areas who work now in urban settings due to long hours of working, poor housing conditions, insufficient food and rest, have been observed to suffer from tuberculosis. Moreover, these migrants in urban slums, having conflicting patterns of behaviour in cities, get alienated in the home environment. This leads to high rates of drug abuse, alcohol consumption, sexually transmitted diseases (STDs) and AIDS as well as certain mental disorders. The psycho-social stresses also result in an increase in suicide rates, social maladjustment due to the ageing process, disintegration of the aged within the community, spread of alcohol and drug abuse and increase in crime and juvenile delinquency. Further, it is observed that urban malaria has surfaced in many slum regions of cities due to lack of poor sanitation because of open sewers, location of solid waste disposal, pollution of bodies of water and flooding during the monsoon (ibid.).

Presently, pesticide-contaminated food has given rise to serious health hazards as the pesticide residues have been observed to pose the third highest threat to environmentally-induced cancer next only to cigarettes and radiations. Further, road accidents due to increased use of vehicles are getting alarming. It is observed that nearly one-third of the total accidents in the country occur in metropolitan cities of Mumbai, Bengaluru, Delhi, Chennai and Kolkata, and a large proportion of them turn out to be fatal. There are a number of social diseases and addictions such as STDs, tobacco hazards and alcohol poisoning. The prevalence of these is mostly in urban slums. We have also a higher incidence of increase in occupational accidents in the case of workers employed in all manufacturing industries/factories, mines, railways and ports, besides, those working in chemical manufacturing industries.

Air pollution due to industrial expansion has resulted in many kinds of respiratory ailments, especially among the women and children in terms of reduction in birth weight and increased pre-natal death rates.

Health Hazards in the Milieu of Industrialization and Globalization

Industrialization and consequent urbanization of the developing nations, including India, is basically based on the transfer of Western technology into these nations without provision of adequate checks on the emission of pollution caused by energy combustion or burning of chemical substances. The resultant effect has been a rise in the prevalence and incidence of environmental degradation and damage causing many kinds of health ailments and diseases. Research evidence suggests association of social class and economic cycles, social organizations, work environment, production process and environmental pollution to illness and disease (Waitzkin, 1983; Navarro, 1986). It is observed by Bolaria and Bolaria (1994) that in any reference to environmental damage and workplace hazards, chemical industry is critical and its influence needs to be analysed as it is estimated that there are presently thousands of chemicals available in the market. Their use is generating as much as 50,000 tonnes of hazardous waste every year. For example, in the United States of America, it is reported that over one million pounds of synthetic organic chemicals are produced every day and these are released into the environment. Besides these chemical pollutants, new substances also get added every year. Many of these chemicals get added to certain products used in homes like, for example detergents, in construction and insulation work, fertilizers, plastics, cloth fibres, etc. These get into the environment and people are exposed to many of these toxic substances. Even in agricultural operations, there is a dramatic increase in the use of pesticides and herbicides. These chemicals get into surface water and seep into the soil and ground water. Besides, there are spills, accidents and disposal of toxic wastes through which chemicals are released into the environment. Apart from chemicals causing environmental pollution, waste from industrial production, automobile emissions, smelters and aerosols, contributes towards air pollution. Acid rain is another serious air pollution problem. These can cause damage to forests and agriculture and also steel as well as concrete structures. Further, the green house effect and ultra-violet radiations have been observed to be more dangerous than acid rain. In short, acid rain and the associated green house effect and ultraviolet radiations are threatening to the environment and may cause damage to forests, agriculture, birds and human life.

These industries and chemical products which cause air pollution also get into the water, thus threatening the aquatic life due to toxic chemicals, industrial effluents and municipal sewage, resulting in depletion of oxygen supply in water. Besides, untreated sewage is a critical source of bacterial and viral contamination. Apart from municipal sewage, many industries make rivers, streams, lakes and oceans as their dumping stations. Even shipments of hazardous and toxic substances and oil tankers always have potential risks of accidents and spillage.

The expansion of the nuclear industry has put many people at risk of radiation and exposure, especially due to the fallout from nuclear weapons, nuclear accidents, radioactive spills and nuclear waste, etc. This has been a serious matter of concern for the managers of nuclear power plants as their location in certain areas has resulted in health hazards to workers and the community from radioactive spills. Further, it is observed that uranium mining also poses serious environmental and occupational health risks as spills from the mines contain a number of hazardous substances such as contaminated water, diesel fuel, mill tailings and sulphuric acid. In addition to these hazardous spills, the accidents in nuclear power plants pose the most serious danger to the environment. A significant environmental problem associated with the nuclear industry is reported to be disposal of radioactive waste. We have tonnes of this waste produced every year by nuclear power plants and its safe disposal and storage is a serious environmental issue. In brief, we can indicate that chemical air, water and radiation contamination are the main contributors to environmental pollution and degradation in industrialized and industrializing nations. Especially, it hits the health status of those who have poor social and material conditions of living (ibid.).

Further, many of these chemicals have an adverse ecological and human health effect such as cancer, especially melanoma skin cancer, the production of cataracts and depression of the immune system. Mercury poisoning in water is dangerous both for aquatic and human life as it causes neurological damage, kidney and liver dysfunction and death. Many fish that are affected get killed and those that survive become unfit for human consumption.

Among children, it has been observed that exposure to excess lead results in stunted growth and may cause mental retardation, birth abnormalities and may also be a major source of hypertension, stroke and heart attack. Even lead poisoning is associated with poor grades and coordination, high absenteeism from school and higher dropout rate among the affected children. Excessive exposure to radiation has been positively related to illnesses such as cancer, leukemia and genetic changes. In short, there is an increasing relationship established between environmental conditions and health status. Though there is an effect of environmental condition on the health of the whole population, but those who are socially and economically poor are the worst sufferers of environmental degradation. This has been confirmed by epidemiological data which clearly brings out the differential health status. Even in developed countries such as Canada, USA, etc., social and material conditions such as poor housing, poor nutrition, personal and social pathologies contribute significantly to high morbidity and mortality in the low income groups. The same holds good even for the developing nations like ours. Apart from physical ailments, persons from such groups are subjected to psychological illnesses because of their employment condition as they mostly hold low status jobs and have no control over their work conditions, causing dissatisfaction and stress. Further, it is observed that residents of poor neighbourhoods experience higher mortality rates living in rich neighbourhoods. Research evidence suggests that children exposed to air and chemical pollution and noise show blood pressure abnormalities and certain cognitive and behavioural difficulties (Starfield, 1984).

A study done in the United States also shows that lead poisoning is far more common in minority and low-income children who are more exposed to chips of lead-based paint and other sources of lead poisoning such as gasoline (Mahaffey et al., 1982). It is also reported that poorer the area, the shorter the life expectancy of both men and women residents. For example, the mortality for lung and oral cancer for men and lung and cervical cancer for women increases as income decreases (National Cancer Institute, 1990). Further, women and racial minority workers in some sectors have disproportionate burden of work-related and environmentally-produced ill-health. We have evidence to support that farm workers, garment workers and female domestic workers and cleaners are likely to be exposed to numerous health hazards largely due to unsafe and insanitary living conditions, exposure to herbicides and pesticides and arduous tasks of textile and garment workers in 'sweat shops'. Both the textile workforce and the domestic cleaners are mostly composed of racial and ethnic minority women and they are likely to be subjected to physical as well as psychological health risks (Bolaria and Bolaria, 1994).

Notwithstanding the occurrence of many diseases as a consequence of environmental pollution in the developed nations, there are even visible effects of these in developing nations like ours. We have, in India, occupational accidents and injuries in the workplace environment due to introduction of new technology in the manufacturing industries. For example, it is observed that out of 100 workers employed in all manufacturing industries, nearly 60 of them are subjected to various kinds of accidents whereas in UK, it is nearly 34 and in USA, it is nearly 25 per cent. It is roughly estimated that the number of people killed due to occupational injuries would be in the region of 1.5 lakh per year whereas the number of permanently disabled and seriously injured may be in the region of 15,000–30,000. Besides, the number of workers killed in manufacturing and mining industry is estimated to be 6,000–18,000 per year while in the agricultural sector, around one lakh per year and nearly 10 to 20 lakh have non-fatal injuries. Lack of skill, empathy in attentiveness and proneness to accidents are usually considered to be the major causes of these accidents. Further, production relations, work schedule, remuneration and job satisfaction, etc., also contribute towards that end. Even national economic structure, a large pool of unemployed labour, global economic competition and state policies are significant contributing factors (Harris, 1992).

The other hazardous industry causing environmental health problems in our setting is the chemical industry. This industry nearly contributes 7 per cent to the GNP (Gross National Product), which may be roughly half of the industrial outset and implies 6 per cent of the industrial labour force. As in 1987–88, exports were to the tune of ₹ 13,000 crore while having a significant annual growth rate. It is observed that nearly 5 million chemicals got synthesized in the last 40 years at the rate of 10,000 new ones each month. Because of the chemical pollution which has caused ground water contamination, there are personal and environmental hygiene problems having short or long-term repercussions on the life of human beings. The toxic hazards have been observed extensively in our country because of poor legislation and implementing machinery, whether it is chemical pollution or air pollution due to thermal power station or sewerage disposal and treatment or even industrial effluents. The State Pollution Boards that have been set up in each of the states have failed to provide adequate control mechanisms to check these. We need to evolve a sound environmental policy and establish adequate health infrastructure for having minimal conditions conducive to people's health and productivity so as to reduce the destructive practices. To realize that, there is an urgent need to have state private and NGO partnership (ibid.).

Recently, it has been observed that the seepage of arsenal in the ground water due to heavy inputs of chemical fertilizers is causing cancer among some segments of the population in certain areas of Punjab, especially those under intensive cotton crop coverage. There are even reports that the state government in Punjab is planning to shut down a thermal power station located in Bathinda as it is causing severe environmental health problems for the people living around. Besides, there are a number of reports which have shown that in chemical industries around Udaipur, Kanpur, Delhi, Durgapur and industrial units such as Durgapur State Quarry, East India Pharmaceutical Works, Durgapur Chemical, Alloy, Steel plant, Hindustan Fertilizers Corporation Ltd., Phillips Carbon Black Ltd. and Durgapur Steel Plant, etc., the suspended particulate matter has been observed to be higher than the permissible limit both in respect of air emissions and water effluents.

It is also observed that some of the major cities in India such as Mumbai (Thane), Kolkata (Howrah), Ahmedabad, Bhopal, Trivandrum, Hubli-Darwar in Karnataka, Vishakhapatnam, Jaipur, Delhi, Shimla, Lucknow, Patna,

Bhubaneswar, Guwahati, etc., lack a proper sewage system and due to intensive industrialization of some of these cities, the Environment Pollution Control Boards have not been able to maintain the permissible limits of pollution. This has led to both water and air pollution causing many kinds of serious diseases, especially mosquito-borne diseases and tuberculosis among the slum dwellers (TB, asthma, bronchitis, cholera, dysentery and skin diseases). It is reported that cholera, dysentery, gastroenteritis and skin diseases are on the rise in urban areas and they account for nearly 60 per cent of all urban deaths (Sheth, 1997). Even in north-west India, due to industrialization in growing cities like Faridabad, Gurgaon, Panipat, Yamunanagar, Ludhiana, Jalandhar and Amritsar, the environmental pollution has caused water, air and hygiene related diseases.

The global reach of the multinational corporations (MNCs) as a result of globalization process in the post-reform period of 1991, due to expansion of industrial production across the country, has posed health hazards not only to the workforce and people living around, but also to population living far away. For example, the Chernobyl nuclear plant accident in the former USSR and the Union Carbide tragedy in Bhopal had ramifications faraway too. The nuclear accident in Chernobyl was found to have effects on agricultural products throughout much of Europe and the Bhopal gas tragedy also had a far-reaching influence on the population surrounding the plant. Further, it is observed that acid rain and nuclear accidents do not limit themselves to national boundaries. The MNCs, in order to have profits, generally look for areas having low production cost, cheap labour, diversification possibilities, access to cheap resources and new markets. As such, they invest in the developing countries as the business climate in these nations is very favourable for profits and capital accumulation. In addition, they can avoid rigid environmental standards of health and safety regulations regarding hazardous industries. It is reported that millions of workers in the Third World countries employed in multinational corporations draw poor wages and are exposed to hazardous work conditions (Bolaria, 1988).

Further, it is observed that nearly one million people are employed in industrial trade zones in the Third World, besides millions more who work outside the zones in MNC-controlled plants and domestically owned sub-contracting factories. It is reported that in light assembly work, 80 to 90 per cent of the workers are women who work for lower wages than men and are preferred because MNCs require a workforce that is docile and easily manipulated.

The workers in the textile industry are also subjected to many health risks such as 'brown lung'. They are also exposed to chemicals in the processing and dyeing of fabrics. Even in the electronic industry, which is perceived as a 'clean' industry and has majority of women employees, health problems such as dizziness, nausea and headaches are common because of chemicals used in the electronics industry. Even in the asbestos industry as observed by Castleman (1981), with regard to the Shree Digvijay Cement Company in Ahmedabad and Hindustan Ferodo in Mumbai (both associated with Western multinationals), workers and those around these plants are exposed to health hazards that would not have been allowed in the West. It is observed that untreated waste litters the entire area surrounding the plant. It is vouched that responsible members of the business should make a commitment not to profit by such abuses. The Third World countries are also becoming dumping grounds for hazardous wastes because of the dangers of toxic chemical substances.

Willing to do repetitive, boring assembly work, women are considered by MNCs as perfect employees with natural patience and manual dexterity (Fuentes and Ehrenreich, 1983, as quoted by Bolaria and Bolaria, 1994). Even the workers in the Third World are also exploited, especially women and children, by having long working hours and compulsory overtime for which they are not paid. They are forced to join for wage concessions while doing labour in an unhealthy work environment as well as live in a polluted environment to maintain their jobs. As stated earlier, there is a tendency on the part of these MNCs to move into those areas which do not demand costly workplace safety and health standards and environmental pollution control regulations. The governments of the underdeveloped countries in many cases are also willing to provide relaxed environmental, health and safety standards and other incentives to attract these industries and jobs. (Health problems among the workers have been observed to vary from eye problems and muscular and skeletal disorders to hand injuries and loss of body parts). Further, it is observed that the pharmaceutical companies also promote medical products that are banned in the industrialized countries with rhetoric in the developing nations of their effectiveness and minimal warning of the dangers of serious or lethal side effects. The activities of the food industry also match that of the pharmaceuticals. For example, the promotion of the infant formula led to 'commerciogenic malnutrition', which resulted in increased infant morbidity and mortality in the poor nations. Already, in the developing nations, there is poverty, insanitary living conditions and low nutritional levels which result in lower resistance to disease and contribute to ill-health and high morbidity and mortality. The additional imported hazards as a consequence of a globalized economy may prove to be more serious to a vulnerable population such as ours as was observed in the case of Bhopal gas tragedy.

Interestingly, it is observed that though people in the Third World are victims of pesticides banned in advanced countries, they are not free from the ill effects of these. No doubt, the 'circle of poison' and the 'circle of victims' start at the point of production because the production workers become the first victims of the circle of poison. But subsequently, these pesticides get back to the advanced countries in the form of imported food, thus completing the circle, in the sense that pesticides add up on breakfast and dinner tables in the form of imported food and other commodities. It is observed that these pesticides are primarily used on crops produced for export purposes in the Third World nations. Food and other commodities such as bananas, coffee, sugar, tomatoes, tea, cocoa, strawberries, pepper and olives are sprayed with pesticides banned in the developed countries, to which these products are exported. Though these food products are produced in poor countries affecting health and lives of the poor and hungry, these are consumed by well-to-do and well-fed people of the developed nations (Weir and Schapiro, 1981, as quoted by Bolaria and Bolaria, 1994).

It may be inappropriate if we do not highlight the Bhopal gas tragedy in our context because it was a severe environmental disaster caused by a multinational (Union Carbide Corporation, US-based MNC) operating a huge pesticide plant on the outskirt of Bhopal city. It was on December 3, 1984 that an accident took place due to flow of water into the toxic methyl isocyanate storage tank. This caused a powerful and uncontrollable reaction as a result of generation of heat. The deadly methyl isocyanate escaped from the tank in the form of a gas which killed thousands of human beings, blinded many more and caused severe damage to the people living nearby in different ways. The Union Carbide plant was located in an area which was a slum dwelling where a large mass of ecological refugees were living. It is reported that within minutes of the gas release, about three thousand people died on the spot and about 50,000 sustained lung damage. This tragedy was considered as an episode of the poisoning of environment in human history by a multinational company. Besides health hazards caused by it, this tragedy also raised issues of environmental prudence, human rights issues and MNC and India State relationships by regional, national and international NGOs, jurist and environmental human right activists who brought pressure on the initial hesitant state and union governments. The case is yet to be solved because of complicated and contradictory statements, facts, evidences, medical arguments, lobbying by powerful interest groups and public-spirited NGOs However, this has created a sound ecological awareness and the subsequent power projects by MNCs (Enron and Congentrix) and other chemical projects have been subjected to careful scrutiny and critical examination by the environmentalists in India (Sheth, 1997).

Preventive and Safety Measures in Health Maintenance and Restoration

In view of increasing industrialization as a result of modernizing and globalizing influences, there is a rise in environment health hazards across the country. This makes sustainable development a difficult proposition. How do we go about combatting deadly diseases which have been caused due to excessive fuel burning, plant emissions and contamination of ground water due to intensive use of chemical fertilizers for raising the agricultural production? Should it be the responsibility of an individual or that of a society through state intervention to maintain or restore the health of the people? Before we discuss this medical ethical issue, it may be apt to have a synoptic view of a historical development of medicine in our country.

In the pre-independence period in the ancient times, during monolithic and neolithic culture, we had Shamans of medicine and folk medicine practices, respectively. The practices were passed on to the next generation. In the pre-historic period, animistic religion got evolved and diseases were mainly attributed to supernatural causes or to the wrath of Gods and spirits. People were treated with amulets, charms, medical rites, sacrifices and talisman. During the Vedic period, a mention is made about invocations addressed to the Gods for good health and recovery from ill health. However, it was Buddhism that brought values like healing qualities, mercy and compassion to the field of medicine. It was Rahula, the son of Buddha, who established hospitals both for humans and animals. Ashoka later developed a hospital system on a larger scale and established two chairs at Taxila and Nalanda University in medicine and surgery, respectively. The Unani system got established during the Muslim period (1,000–1,500 A.D.). But, in the Vedic period between 900 and 1,000 A.D., there was no medical development and as a reaction to Buddhism, superstitions and quackery flourished.

Portuguese are reported to be the first ones to have introduced modern medicine or Western medicine in India. A hospital was established in Goa in 1687. Later on, the French and British brought in Western medicine in the East Coast and started hospitals at the state capitals and district headquarters. Medical colleges were also established in 1835 at Madras, Bombay and Calcutta. The rural areas were largely served by quacks, hakims and vaids, ayurveda practictioners, unani and naturopaths, besides astrologers, priests and magicians, sadhus, unregistered practitioners, barbers, etc., through various traditional healing practices, including charms and sacrifices. But in 1920s, Licentiates of Medicine were encouraged to settle down in rural areas to manage the dispensaries established there. Just before Independence, the Bhore Committee was set up in 1943 which in its report in 1946 brought out a wide gap between the urban and rural health services and emphasized that the rural health services need to be strengthened by having quality practitioners to provide health care to the people. The state should be responsible for providing proper medical care and no individual should be deprived of it even if he or she is not in a position to afford it. Later on, in independent India, the Mudaliar Committee (1961) was not in favour of short-term courses for these positions, like the one of Licentiate degree holders in Medicine during the British period. It was argued by the professional body of physicians that such short courses would lower the standard of medical education in the country and we need to develop medicine in the international perspective (Rao, 1972).

To meet the health needs of people in the rural areas, it was suggested to have a large number of paramedical personnel after suitable training, to be posted in the established primary health centres and sub-centres. This led to the establishment of a cadre of health assistants at the community level under the multipurpose health scheme. As on March 2009, we have 23,391 primary health centres and 1,45,894 sub-centres, whereas there are 4,510 community health centres established in the country. Over the years, with the assistance of the international health organizations such as WHO, UNICEF and World Bank, we have made tremendous progress in building up health and medical care infrastructure throughout the country by establishing specialized medical college based hospitals, postgraduate institutes of excellence such as PGI and AIIMS, hospitals in the government as well as nursing homes in the private sector in urban centres, district hospitals, urban and rural community health centres and dispensaries, primary health centres and sub-centres in the rural areas and producing huge health manpower resource belonging to medical, nursing and paramedical categories. These achievements for a country of continental size like ours are by no means small. But they cannot be compared to the developed world. We have achieved success in combating many of the deadly diseases such as small pox, tuberculosis, malaria, etc. Further, we observe a decline of death rate, an increase in the lifespan of Indian population. For example, the infant mortality rate, which is the most important and sensitive index of health, was 110 in 1981. But now, as per the sample registration system bulletin (October 2009), it is reported to be 53 (58 for rural areas and 36 for urban areas). The crude birth rate is reported to be 22.8 (24.4 for rural and 18.8 for urban population) while crude death rate is 7.4 (8.0 for rural and 5.9 for urban population).

Despite an increase in the number of medical colleges and for preparing professional positions and medical institutes in the indigenous system of medicine, there is a need to restructure medical education and to train the health personnel of different categories to make them community health oriented. Though the number of doctors at PHCs increased from 20,308 in 2005 to 23,982 in 2009, there is still 16.2 per cent shortfall of doctors in the total requirement at PHCs. Further, a shortfall of 69 per cent surgeons, 68 per cent obstetricians and gynecologists, 76 per cent physicians and 73 per cent pediatricians is reported because of increase in the number of community health centres. Overall, there is a shortage of 68 per cent specialist physicians or doctors (Government of India, 2010).

The situation is considered inadequate for providing medical facilities in rural India and bringing about rapid growth in medical tourism, which is likely to go up to 2 billion as per CII-McKinsey report. There is an urgent need to improve medical infrastructure by having more medical colleges. To check the deficit of faculty in the medical colleges, which is reported to range from 25 to 33 per cent as per the Medical Council of India findings, the government has announced creation of 7,000 MD (Doctor of Medicine) seats by 2014 to tide over the crisis. In order to fill these 7,000 additional seats for MD, there is a need to expand medical educational institutions in the country. However, one has to be cautious in regard to maintenance of the quality of education. Recently, the government, under the leadership of Union Health Minister

Ghulam Nabi Azad, encouraged the private companies registered in India to set up medical colleges to augment the human resources in this sector. Besides increasing the retirement age from 58 to 65 years, the Medical Council proposed the age of retirement at 70. As of now, it is reported that the country has the capacity to produce only 35,000 medical graduates annually and there is a need to add 10,000 medical seats each year between 2012 and 2016 to get at least 50,000 medical undergraduates over the next five years. We have 4.5 lakh deaths per year in childbirth alone because of lack of trained obstetricians. Further, we have just one doctor for 1,700 people whereas for China, this ratio is at 1: 1,063, Korea 1: 951, UK 1: 469, US 1: 350, Brazil 1: 844, Singapore 1: 714, Japan 1: 606, Thailand 1: 500 and Germany 1: 296. WHO has recommended one doctor for every 1,000 persons. It may take another 15 years to achieve this goal. That is why the suggestion of the Health Ministry to have a three-year rural health care course may be good to address the huge shortage of doctors in the villages (*The Tribune*, Editorial, August 22, 2011).

We have had two National Health Policies, one in 1983 and the other in 2002, where efforts have been made to have community participation for providing adequate preventive and curative health care both in the urban and rural areas by having health workers and village health guides to decentralize the planning process for health delivery system.

Recently, the UPA government under the Common Minimum Programme launched a National Rural Health Mission (NRHM) in the budget for 2005–06 to provide adequate finances, well-motivated village health workers and improved managerial skills and accountability to provide affordable, effective primary health care along with family welfare, sanitation and hygiene, nutrition and safe drinking water facilities. This is to be achieved by having accredited social health activists (ASHA, a cadre of female health volunteers) to strengthen the health services at various levels.

All such efforts to involve the community are laudable but because of the factious or faction-ridden nature of community, it becomes non-viable and impractical. Moreover, it is observed that the decentralization in the planning process within the centralized system of administration often results in being dysfunctional. For effective primary health care at the community level, there is a need to develop a strong referral system for incurable diseases for treatment at the hospital level. Further, we can have health co-operatives at the village level by having contributions from villagers according to their economic status and provision of medicines at 'no profit and no loss basis' (Mehta, 1992).

Despite expansion of medical services, we still observe that communicable diseases take a heavy toll of people, especially in rural areas, due to the prevailing insanitary conditions and non-availability of safe drinking water. Malnutrition, especially among women and children, is promoting diseases such as blindness, goitre, mental disorders, anaemia, etc. Further, the prevalence of cancer, STDs (sexually transmitted diseases), AIDS (HIV positive cases), leprosy and accidents has increased over the years. But, we do observe a large drop in leprosy cases. The services provided at the public hospitals are not satisfactory. The indifferent attitude of physicians, rude and arrogant behaviour of nurses and paramedical staff and unsympathetic behaviour towards patients have been identified as critical factors for a low level of patient satisfaction in the hospitals. It is argued that we already have a huge health resource (say 3 to 5 lakh) of indigenous medicine practitioners like Ayurvaids, Homeopaths and Unanis, etc., in the rural areas who have a well-established rapport with the community. In view of doctors' shortage as well as their reluctance to go to rural areas for want of facilities, it may be worthwhile to induct them in the main health care system by training them in allopathic medicines for 3–5 years on a long, continued basis through short courses of 3–6 months' duration. There is also an urgent need to have intermediate-level courses in medicine for 2 years after graduation either in science or nursing, as most of the physicians after 5–7 years of medical education become urban-oriented and are reluctant to go to rural areas (ibid.).

This suggestion of having short medical courses of three years duration to prepare medical graduates was recently taken up by Health Minister Ghulam Nabi Azad, but the medical fraternity has once again opposed it because of their belief that this will lower medical standards below the international level.

However, the medical fraternity must realize that the ground reality of shortage of availability of medical doctors as perceived by the political system needs to be translated into action by preparing physicians through short-term courses.

A recent study on the operation of the NRHM in Tamil Nadu has brought out that there is some improvement in providing primary and maternal health care services, especially for women and newborn babies, as observed from the increasing patient inflow and the opinion expressed by community members, but they are yet to be close to Indian public health standards. One significant finding is that those primary health centres which showed successful results had good leadership of Deputy Director of Health Services (DDHS) and the medical officers of the PHCs. This also holds good for involving panchayats in the National Rural Health Mission (Palanithurai and Ramesh, 2011).

We can develop confidence in semi-professional nurses or midwives or other para professionals if we imbibe a team spirit among all of them by the team leadership provided by physicians, especially at the PHCs. For that to happen, the authority structure in the health set-up has to become less rigid and acquire a somewhat positive response towards the subordinates (Mehta, 1992).

In the wake of what has been achieved in creating a huge health infrastructure across the country and the desirability of building a transparent and accountable health care system by streamlining the curative as well as preventive and promotive health care, it may be worthwhile to highlight as to whose responsibility is healthcare? Is it that of an individual or a group or society through the intervention of the state?

It is observed that health is as much the responsibility of an individual as that of the society. Further, it is argued that an individual has far greater responsibility in preventing and curing such diseases which may arise from smoking, alcoholism, drug addiction, accident, mental disorders and deviant sexual behaviour. An individual is also to take precautions in promoting public health measures such as waste disposal or safe drinking water availability, etc. In fact, the nature of disease or health hazard may determine the responsibility of an individual or society over health. In addition to preventive and rehabilitative health behaviour, the individual patients should also cooperate with the physicians to receive curative treatment (ibid.).

In other words, an individual should have necessary motivation to seek medical care or he should have enough sensitization to inculcate desirable health practices in his/her behaviour pattern through the use of both personal and impersonal channels of communication. Through health education, his fears, doubts, apprehensions and misconceptions need to be removed for building up positive attitudes towards health. Apart from an individual, the family, especially for the maintenance of a healthy home environment, has a significant role to play in matters related to health as infant, child and maternal health need to be taken care of at this level. Mothers' health values and mothering skills are important for the development of the young. In addition, nutrition is a key factor in maintaining health and we have to take into account the cultural norms related to preparation of food and eating behaviour in the family. It is recognized by the medical fraternity that the traditional practice of breast-feeding is good for infants and provides sufficient feeding to them.

There are also certain cultural patterns which affect child rearing, family and social aspirations competitiveness and weakening of social solidarity. This has resulted in having a high incidence of psychological diseases, suicide and violence. Further, group norms concerning smoking, drinking, sexual practices and nutritional intake either predispose or may even protect persons in the family from the risks of disease. Even the decision to take a particular type of medicine is determined in villages by the authority pattern operating within the family.

In addition to the family, the community, whether urban or rural, has to assume an important responsibility over health matters related to environmental protection. Public health measures, sanitation, safe drinking water supply, disposal of right soil or excreta, immunization of children against diseases, maternal and child health care, nutritious food and health education of community members can only be achieved by participation of people in these activities.

Although the health or medical care system has to be provided by society through state interventions, yet co-operation and participation of people in the health measures is necessary for restoring and maintaining health. The state must take upon itself the responsibility of organization of the health or medical care system by the generation and allocation of resources, training of health personnel at different levels, availability of drugs and medicine, supply of equipment, creation of infrastructure such as hospitals and dispensaries at different levels and above all, a political will to implement health plans and policies. In order to arrive at societal health needs, there is an effort needed to assess the health requirements at the individual, family or group level.

In short, it is suggested that health may be conceived as having two major components. One, maintenance while the other as restoration. For health maintenance or to remain healthy, an individual has a far greater responsibility than the society. However, the health care system should include in its approach a component of health education to modify the health behaviour of individuals to inculcate in them desirable health practices relevant to and determined by the culture of the society. However, if the health needs of a group, community or society necessitate health restoration more than health maintenance, then the society has to assume a greater responsibility over health by providing adequate and suitable health and medical delivery services at various levels within the nation-state set-up. In other words, the society must organize and manage available health knowledge, technology and resources within an acceptable political ideology. This does not exclude the responsibility of society towards health maintenance of its members. In fact, society has to assume the dual responsibility of health restoration and maintenance, especially in the context of developing nations where a vast number of people are illiterate, poor, underfed, living in insanitary conditions and where the environment is conducive to spread of chronic and acute forms of diseases. In a nutshell, the problem of 5Ds as identified by White (1968) – death, disease, disability, discomfort and dissatisfaction – can be tackled by intervention of both an individual and a society through suitable methods and techniques related to prevention, diagnosis, treatment, cure and rehabilitation of sick persons (Mehta, 1992).

To conclude, one may highlight a three-pronged approach in providing health maintenance or restoration to the people whether in the rural or urban areas in relevance with their cultural needs. First, we must generate quality health care or medical services and provide these extensively to all in an equitable manner. This may require a larger budgetary provision. In fact, the government is struggling hard to raise the budget allocation of health sector from less than 1 per cent of GDP to even up to 3 per cent. Whenever there is a financial crunch in development plans, the axe always falls on the health sector.

Secondly, we need to evolve suitable legal measures and provide laws to check not only environmental health hazards at the individual, family or community level, but also for certain projects on a grand scale such as power projects, mining, dam construction, nuclear plants, chemical industries, etc. Unfortunately, the implementing agencies of these laws have not been able to check the violation of pollution control mechanisms provided under these laws. We have more than 200 laws already framed for environmental protection of various nature, but these are not being strictly followed.

Thirdly, the most significant approach needed for health restoration and maintenance is health education of all in different strata of our society. We need to evolve suitable health education methods by having a media mix of traditional and modern channels of communication while keeping in view our social structural and cultural perspectives. This must be integrated with the delivery system evolved for providing health and medical care services by all professionals (physicians, semi-professionals (nurses, etc.), and community health workers/voluntary health guides (ASHA) operating at different levels.

In short, these three approaches need to be coordinated at all levels of the delivery system, while these may be suitably modified in the context of different states of our country in relevance with their respective cultural links to make them more effective.

References

Allison, Ken (1992), 'L'education Sanitaire: la responsabilite' individuelle et al victime prise consume bore c 'missane', *Education Sanitaire*, Vol. 20, Nos 3–4, pp. 10–13.

Azad India Foundation, http://social-issues/status-of-sanitation-in-cities-in-india.html, 12 May 2010.

——. 3 May 2010.

Banerji, D. (1982), *Poverty, Class and Health Culture*, New Delhi: Prachi Prakashan, as quoted by Jill Carr-Harris' 'Environment and Health', in *State of India's Health*, Alok Mukhopadhyay (Ed.), New Delhi: Voluntary Health of India, 1992.

Becker, Howard S. (1963), *Outsiders: Studies in the Sociology of Deviance*, New York: The Free Press.

Berger, Peter C. (1963), *Invitation to Sociology: A Humanistic Perspective*, New York: Anchor Books.

Bolaria, B. Singh (1988), 'Profits and Illness Exporting Health Hazards to the Third World', in *Sociology of Healthcare in Canada*, B. Singh Bolaria and Harley D. Dickinson (Eds), Toronto: Harcourt Brace Jovanovich, pp. 477–496.

——. (1988), 'The Health Effects of Powerlessness: Women and Racial Minority Immigrant Workers', in *Sociology of Healthcare in Canada*, B. Singh Bolaria and Harley D. Dickinson (Eds), Toronto: Harcourt Brace Jovanovich, pp. 439–459.

Bolaria, B. Singh and Rosemary Bolaria (1994), 'Globalization of Environmental and Industrial Health Hazards', in *Racial Minorities, Medicine and Health*, B.S. Bolaria and Rosemary Bolaria (Eds), Halifax, Nova Scotia: Fernwood Publishing.

——. (1994), 'Inequality and Differential Health Risks of Environmental Degradation', in *Racial Minorities, Medicine and Health*, B.S. Bolaria and Rosemary Bolaria (Eds), Halifax, Nova Scotia: Fernwood Publishing.

Brown, E. Richard and Glen E. Margo (1978), 'Health Education: Can the Reformers be Reformed?', *International Journal of Health Services*, Vol. 8, No. 1, pp. 3–25.

Brown, N.N. et al. (1974), 'Life Events and Psychiatric Disorders: Some Methodological Issues', *Psychological Medicine*, Vol. 3.

Castleman, Barry I. (1981), 'Double Standards: Asbestos in India', *New Scientist*, 26 February, pp. 522–523.

Cockerham, William C. (1978), *Medical Sociology*, Englewood Cliffs, NJ: Prentice Hall.

Dingwall, Robert (1976), *Aspects of Illness*, London: Martin Robertson and Company Ltd.

Dubos, Rene (1969), *Man, Medicine and Environment*, New York: Mentor.

Freeman, Victor J. (1970), 'Human Aspects of Health and Illness: Beyond the Germ Theory', *Journal of Health and Human Behaviour*, Vol. 1.

Freidson, Eliot (1970), *Profession of Medicine*, New York: Dodd, Mead and Company.

Freudenberg, Nicholas (1978), 'Shaping the Future of Health Education: From Behaviour Change to Social Change', *Health Education Monograph*, Vol. 6, No. 4, pp. 372–377.

Fuentes, Annette and Barbara Ehrenteich (1983), *Women in the Global Factory*, New York: South End Press, as quoted in 'Globalization of Environmental and Industrial Health Hazards', in *Racial Minorities, Medicine and Health*, B.S. Bolaria and Rosemary Bolaria (Eds), Halifax, Nova Scotia: Fernwood Publishing, 1994.

Government of India (July 2010), *Rural Health Statistics in India* (March 2009), New Delhi: Ministry of Health and Family Welfare.

Harris, Jill-Carr (1992), 'Environment and Health', in *State of India's Health*, Alok Mukhopadhyay (Ed.), New Delhi: Voluntary Health Association of India, pp. 87–128.

Kasl, S.V. and S. Cobb (1966), 'Health Behaviour, Illness Behaviour and Sick-role Behaviour', *Archives of Environmental Health*, Vol. 12, pp. 246–266, 531–541.

Kosa, John and Leon Robertson (1969), 'The Social Aspect of Health and Illness', in John Kosa et al. (Eds), *Poverty and Health*, Cambridge, Mass.: Harvard Press.

Lipovski, Z.J. (1970), 'Physical Illness, the Individual and the Coping Process', *Psychiatry in Medicine*, April issue, pp. 91-100.

Mahaffey, Kathryn R. et al. (1982), 'National Estimates of Blood Lead Levels in United States 1976–1980', *New England Journal of Medicine*, Vol. 307, No. 10, pp. 573–579.

Mckinley, John B. (1975), 'A Care for Refocusing Upstream: The Political Economy of Illness', in Applying Behavioural Sciences to Cardiovascular Risks, Proceedings of the American Heart Association Conference, Seattle.

Mechanic, David (1978). *Medical Sociology* (Second Edition), New York: The Free Press.

Mehta, S.R. (1992), *Society and Health: A Sociological Perspective*, New Delhi: Vikas Publishing House Pvt. Ltd.

National Aids Control Organization Report (2008), New Delhi: Ministry of Health and Family Welfare.

National Cancer Institute (1990), *Canadian Cancer Statistics*, Toronto, as quoted in 'Globalization of Environmental and Industrial Health Hazards', in *Racial Minorities, Medicine and Health*, B.S. Bolaria and Rosemary Bolaria (Eds), Halifax, Nova Scotia: Fernwood Publishing, 1994.

Navarro, Vincente (1986), *Crisis, Health and Medicine*, New York: Tavistock Publications.

Palanithurai, G. and R. Ramesh (2011), *Globalization and Rural Development*, New Delhi: Concept Publishing Company Pvt. Ltd.

Parsons, Talcott (1951), *The Social System*, New York: The Free Press.

Penfold, Labonte Ronald (1981), 'Analyse critique des perspective canadiennes en suzan promotion de la santé', *Education Sanitaire*, Vol. 19, Nos 3–4, pp. 4–10.

Rainwater, Lee (1963), 'The Lower Class: Health, Illness and Medical Institutions', in *Among the People*, Irwin Deutscher and Elizabeth Thomson (Eds), New York: Basic Books.

Rao, K.N. (1972), 'Organizational Structure of Medical Practice in India', paper presented at a Seminar on Social Sciences in Health (ICMR-ICSSR), New Delhi: Central Health Education Bureau (Mimeograph).

Rosenstock, Irwin, W. (1966) "Why People Use Health Services?", *Milbank Memorial Fund Quarterly*, 64, pp. 94–127.

Sheth, Pravin (1997), *Environmentalism: Politics, Ecology and Development*, New Delhi and Jaipur: Rawat Publications.

Starfield, Barbara (1984), 'Social Factors in Child Health', in *Ambulatory Paediatrics*, Morris Green and Robert J. Haggerty (Eds), Philadelphia: W.B. Saunders, pp. 12–18.

Strauss, Anselm (1969), 'Medical Organization, Medical Care and Lower Income Groups', *Social Sciences and Medicine*, Vol. 3, pp. 143–177.

The Tribune (August 22, 2011), 'Augmenting Infrastructure: Medical Education Needs a Boost', Editorial, Chandigarh.

Waitzkin, Howard (1983), *The Second Sickness: Contradictions of Capitalist Health Care*, New York: The Free Press.

Weir, David and Mark Schapiro (1981), *Circle of Poison: Pesticides and People in a Hungry World*, San Francisco: Institute for Food and Development Policy, as quoted in 'Globalization of Environmental and Industrial Health Hazards', in *Racial Minorities, Medicine and Health*, B.S. Bolaria and Rosemary Bolaria (Eds), Halifax, Nova Scotia: Fernwood Publishing, 1994.

White, Kerr L. (1968), 'Organization and Delivery of Personal Health Services: Public Policy Issues', *Milbank Memorial Fund Quarterly*, Vol. 46, No. 1, pp. 225–258.

World Health Organization Report (July 2011), 'Sick Water', gits4u.com.

Zola, Irving K. (1965), 'Illness Behaviour of the Working Class', in Arthur Shostak and William Ginberg (Eds), *Blue Collar World*, Englewood Cliffs, NJ: Prentice Hall.

4

Population Growth, Environmental Degradation and Sustainable Development

Population Growth: A Demographic Perspective

Of all the species, man is considered to be a unique human being because he has the capacity to think about his environment as a detached independent observer. Simultaneously, his activities on planet Earth have modified natural cycles and changed the environment on a scale which is greater than that of any other animal. It is observed that the present world population is acquiring dangerous proportions and if we grow at the present rate, it may not be possible for Earth to support them. As such, there is a dire need expressed to limit population growth so as to take care of private and personal activities concerning social, educational and economic concerns of human beings (Harvey and Hallett, 1977).

Historically speaking, the world population grew very slowly from about 2.5 million at the beginning of urbanization to around 15 million at the time of black plague of the middle ages. However, as a result of industrial, scientific and medical revolution and the colonial expansion of Western powers, the world population has touched dizzy heights. It was during the 20th century that the world's population increased nearly fourfold, that is from 1.6 to 6 billion. In the next century (21st century), the world population may acquire a gigantic number of 12 billion people and that may reduce areas on the earth to preserve wildlife. Besides, that may put extreme pressure to raise food production and have water and non-renewable resources (Fielder and King, 2010).

Population and Food

The crucial problem for human existence on the planet is food security. There is a need to have sufficient food to maintain a healthy and productive life of human beings by having efficient and low-cost food systems compatible with the sustainable use of natural resources. This is what Vision 2020 of the International Food Policy Research Institute has indicated. It is expressed that human population is increasing exponentially (growth at a fixed rate of compound interest), which is a common feature of the biological system whereas the food production is going up only in a linear fashion. It is projected that the world population will increase from the current 6.96 billion to around 10 billion by 2050 and 80 per cent of it is expected to be in the developing countries. An estimated one billion people or 14 per cent of the world population is still chronically malnourished and most reside in the developing nations. This necessitates doubling of the food production in the next 30 years and triple the quantity in the next 50 years. No doubt, the cereal yields, like population, doubled during 1960–2000, yet the average yields of root and tuber crops increased by only 36 per cent and that of pulses, only by 29 per cent during the same period. As such, the Green Revolution was promising for cereals but not for tuber and pulse crops (Kang, 2011).

In the Indian context, the population has risen from 370 million in 1950 to 1,189 million now and is expected to increase to 1,657 million by 2050. This gives an idea of an increase of nearly 40 per cent over the current population. However, it is expected that the population growth, which is presently around 1.2 per cent, may decrease to 0.4 per cent by 2050. The disturbing factor is that India's arable land has been shrinking for a long time. (In 1979, it was 163.4 million hectares but in 2009, it got reduced to 158 million hectares, which is a decrease of 5.4 million hectares). Even the investment in agricultural research and development has been on the decline during the last few decades. (Our investment in overall R&D effort is nearly 6 dollars per capita, which is very small as compared to that of China, which is 12 dollars per capita, and for the United States, it is 810 dollars per capita). Our annual food production at the current rate is about 235 million tonnes, but we may need around 280 million tones by 2021 and around 450 million tonnes by 2050. Moreover, there is depletion of water resources in the Indo-Gangetic plains where most of the rice and wheat are produced. We, in India, need to increase our outlay for Agricultural Research and Development and to provide a check on the conversion of farmlands for building and other non-agricultural purposes. Further, there is a need to evolve a policy on development and use of natural resources. Besides, there is a need to check food wastage and enhance efficiency of water use for agricultural purposes. FAO expert Alexander Mueller has pointed out that only 50 per cent of the food produced is actually consumed, whereas the other half is lost in storage, distribution and at the level of end-users. As such, India needs not only to check food wastage but will also have to be competitive and innovative in developing new technologies through research and development (ibid.).

Food insecurity due to population growth is becoming a critical factor not only in India but in many other regions of the world. There is a need to bring in new productive areas to feed the unsustainable human populations that are increasing due to our overstressing impingement on soil and ecosystems through application of artificial fertilizers and pesticides. As a result of that, the best productive areas of the planet are slowly getting reduced to marginal lands. Further, some of the best regions which are having major populations are being appropriated for urban and industrial development.

In the absence of long-term sustainable productivity, it is expected that there will be a continued crisis in food production, as populations crest in the early 21st century. Besides, people within their own country may find it difficult to buy food as many landowners, especially the large ones, may divert their efforts to growing more cash crops such as cotton, coffee, tea, sugar or tobacco for export purposes. This is being encouraged by such governments which are over-burdened with debt to earn foreign exchange. It is a matter of debate whether the growth of cash crops has reduced food production or not, but one thing is certain that these tend to take up the best of land while pushing subsistence farmers into areas with poor soil and rainfall, thus reducing the yields on the marginal lands. Further, cash crops also receive most of the Third World's credit, fertilizers and pesticides and advice on agriculture to promote these.

Due to the fall of food production, especially in Africa, more and more of foodgrains had to be imported. However, grain production, which provides about half the world's calories, has increased from about 700 metric tonnes in 1950 to over 1.8 billion metric tonnes in 1986, that is at a rate of 3 per cent a year. No doubt that has outstripped the population growth. Even meat, milk and fish production rose by 2 per cent annually between 1965 and 1986 and the harvest of vegetables, pulses and fruit grew by 2.5 per cent a year. The World Commission on Environment and Development (1987) has attributed this increase mainly to the development of high-yielding new seed varieties, a nine-fold increase in the use of chemical fertilizers, a 34-fold rise in pesticide appreciation and a doubling of the world's irrigated crop land from 135 million hectares in 1960s to 271 million hectares in 1985.

But this agricultural boom may not last for long as it was reported that the global harvest started falling in the second half of the 1980s. The per capita grain production, which grew from 246 kg to 345 kg between 1950 to 1984, came down to 296 kg by 1988, a level that was prevailing in mid-1970s. In 1988, for the first time in history, USA produced less grains than it needed for its own people. One of the major reasons for this fall was attributed to bad weather conditions. For example, there was a drought which hit India in 1987 and the US, Canada and China in 1988, the world's three biggest food producers. Another major cause of falling harvest is over-use of land causing erosion and desertification. The Food and Agriculture Organization of the United Nations estimated that the soil erosion could take 65 per cent of the Third World's rainfed (non-irrigated) land out of production by the year 2000. Further, every year the World may lose 1.5 million hectares of irrigated fields to salinization. As such, the most important priority has to be to increase both production and consumption, especially in the developing countries, as most of them have a tendency to concentrate their resources on the cities and on industry, thus neglecting the farming and the countryside. Even the food prices are kept low to please the city dwellers and this causes ruination of agriculture. This bias must be ended as increasing food prices may benefit the middlemen than the farmers, as was also observed during the Green Revolution. This approach may not help the poor or reduce hunger. As such, we need to focus on small farmers, who constitute the bulk of the poor in many countries and have the potential for raising production. Wherever, for example in Zimbabwe in 1980 after its independence, small farmers were encouraged and given credit, the harvest increased and hunger had fallen. In this regard, land reform is particularly significant as it may result in splitting of big estates which are not intensely farmed. This may give poor farmers and landless people the means to grow food to feed their families. It is observed by the World Bank that such a 'patch work revolution' may increase yields even faster than the Green Revolution and may show more success in reducing hunger (Fielder and King, 2010).

It has been observed that population growth increases the share of population accounted for by agriculture and this may also enhance the proportion of land area allotted to food production. The research on 'Population and Food' has shown a depressing effect of demographic increase in polluting the environment as well as in making many kinds of adjustments to population pressure against the natural resource base. But, unfortunately, we do not have strong grounds for either optimism or pessimism regarding the role of population on food requirements, though most of the deforestation in mid-latitude tropical forests has occurred due to expansion of agriculture (Lutz, 2002).

A latest study on worldwide food insecurity by the UN Food and Agriculture Organization (FAO) estimated that a total of 925 million people were undernourished in 2010 and two-third of them lived in just seven countries such as Bangladesh, China, The Democratic Republic of Congo, Ethiopia, Indonesia and Pakistan. Besides, it is observed that despite the world potential to produce enough food to feed everyone, nearly a billion people (100 crores) go hungry everyday or eat so little that we can consider World Food Day (celebrated every year on 16 October) as Global Lack of Nutrition Day. The main reasons identified for so many people remaining hungry are rising food prices, stock market speculation in crop futures, conflicting climate change and corrupt repressive regimes. It is also because of humanity's inability to grow the crops. Further, it is observed that some 75 per cent of those going hungry are rural people, who derive most of their living from agriculture but unfortunately agriculture has been neglected for too long and we need to invest more in research and rural infrastructure or we will face severe food security crisis in the future. Industrialization alone may not get us out of economic difficulties. As such, it is important to strengthen movements for public awareness on the scarcity of food and mobilize them against hunger, malnutrition and poverty (Dugan and Randall, 2011).

Population and Development: Social and Economic Consequences

For centuries, scholars have debated about the significance of population as a factor in the political economy of nations as well as in economic growth and sustainable development. It is reported that population deals with birth and death, sex and marriage, gender roles, intergenerational relations and with inter-regional migrations. As such, it is conceived as a very sensitive and emotional topic focusing on the foundation of culture, religion and national identity, population size, structure and spatial distribution are significant both at the local and national levels. Further, the increase in globalization of human migration, financial flows, population dependents, consumption and emissions makes it imperative to examine the population variable from a global perspective (Lutz, 2002).

Population has been observed to be a significant factor in development. In short, population change may depend upon the surplus of births and deaths or vice versa, plus the balance of in-migration and out-migration. Generally, a higher birth rate is accompanied by a higher death rate which varies with nutritional standards, sanitary conditions, level of medical knowledge and availability of medical facilities. Twentieth century has observed a decline in the death rate in the Western or developed nations and a similar trend has been observed in respect of birth rate. In line with the transitional theory of population, it is expected that a similar decline of death and birth rates will occur in the developing nations (Mehta, 1997).

We have recently touched the 7-billion mark in population size at the global level. The projections made by the International Institute for Applied Systems Analysis have indicated that global population may not double from its present size but there may be a manifold increase in the size of the population above 60 yeas, which may double or even triple in the 21st century. This study done by Lutz and his associates, which appeared in *New Scientist Journal* (17 February 1996: 8), brought out that for the first time fertility rates seemed to have declined in every region of the world in terms of number of children per family, which may alter further. Besides, there is a possibility that death rates may also change in the future because of changes in the rates of starvation and in the quality of medical care (the study was based on the basis of dividing the world into 13 regions and 12 population experts were asked to predict the most likely rates of fertility and mortality for each region in 2050 and 2100. The experts were asked to choose high and low extremes that were 90 per cent certain to embrace the true values. Their predictions were combined statistically to project the world population). The estimates showed that the population will grow from 5.75 billion in 1996 to 10 billion by 2050, reach a peak of about 11 billion by 2075 and remain almost at that level or may slightly decline towards 2100. There is a 95 per cent probability that the world population will be between 6 and 17 billion in 2100, with 64 per cent chances that it may not reach double its present level.

Historically speaking, global population at the dawn of the agricultural revolution was about 2.5 lakh which in 1800 reached one million mark (equivalent to the present population of Europe and North America combined). It took another 130 years to reach the second billion (1930) and another 30 years until 1960 to reach the third billion mark. The fourth billion was added between 1960 and 1975, while the fifth billion mark was achieved in 1987. We reached the sixth billion number in 1999 and now the seventh billion mark in 2011. Further, it is observed that the total fertility rate, that is TFR, declined modestly in most parts of the world during 1950–55 to 1970–75 period. It further declined over the following 20 years with a rapidity which was unimaginable in the 1960s. This was mainly pronounced in Asia, where TFR fell to more than two children per woman because of dramatic fertility decline in China during 1970s. However, Africa remained an exception; the fertility rates there remained much above six children on an average per woman through the late 1980s. Beyond that, there was some beginning of fertility decline observed. However, regions like Europe and North America, which had already achieved very low fertility by 1970–75, showed the same level or even there further falls were observed. We have evidence to show that during the 1950s–1960s due to spread of modern hygiene and medicine, there was a higher reduction in mortality than the decline in fertility rates (for example, during the 1950–55 period, life expectancy was lowest in Africa at 38 years and Asia at 41 years, while it had improved significantly in Latin America at 51 years. In the next 20 years, the life expectancy increased tremendously in all parts of the world. For example, in Asia, which is the most populous continent of the world, it increased by 15 years over the short period but in Africa it increased by eight years, which was below the world average. Further improvements were observed in Asia, Latin America and Africa over the next 20 years till 1990–95 but at a slower speed). This rapid increase in life expectancy was mainly due to falling of child mortality rates as well as due to better maternal health (Lutz, 2002).

Interestingly, 2002 was observed to be a very critical year when a drop in fertility rates was significantly detected across the diverse countries spanning the developed and the developing world comprising nearly half the world's population. However, there were nothing in common observed between their respective governments and social attitudes. In 1950, worldwide, on an average, women had five children but in 2002, it was observed to be 2.3. This fall in TFR indicated a downgrading of future population predictions. Fears have been expressed of a population crash because societies full of elderly persons are unable to support them from their own services. No doubt, a few of the countries which have shown a decline in population growth, like China, have forced contraception or sterilization on their people, yet in a country like Brazil (which has no state family planning programme because of opposition from the Catholic church), millions of its women attended sterilization clinics and the fertility was almost halved in 20 years (2.3 children per woman in 2002). The case of Iran is somewhat more remarkable. Despite the fact that in 1994, the Mullahs who ruled the country attended the UN Population Conference in Cairo and opposed vehemently the international agenda for cutting birth rates, their women back home took control of their 'bodies' and made the fertility rates crash from 5.5 children per woman in 1988 to just 2.2 in 2000. Another classic case is that of Italy (which is home to the Catholic church and is opposing artificial birth control methods), which is showing super low fertility rates much below the replacement levels. At just 1.2 children born to each Italian woman, the rate is little more than half the number needed to prevent the population growth and that comes close to being matched by Greece, Spain and Czech Republic (Fielder and King, 2010).

However, there are some conservative Islamic countries such as Saudi Arabia, Afghanistan and Pakistan which show the highest birth rates. Even in India, the Muslim community is reported to have grown by 36 per cent between 1991 and 2001 and they account for nearly 13.5 per cent of the total population. On the other hand, Hindus, who constitute nearly 81 per cent of the population, showed a population decline from 25 per cent to 20 per cent in the same period. It is believed that the Muslim community fairs badly in literacy as compared to other groups and that is one reason for their growth. African nations depict uneven decline in the fertility rate but in many of these nations, HIV is wreaking havoc on the young population as 15 million deaths from AIDS are expected in the next five years (as per UN estimates) and majority of them are from Africa. For example, life expectancy in Botswana and Zimbabwe has gone down to 40 years from 60 years.

By and large, female education and employment as well as increasing use of contraception have been observed to show correlation with falling fertility rates (contraceptive use in the developing world has gone up from 1 in 10 couples to more than half of all couples. Further, a 15 per cent increase in the use of contraceptives means one lower birth for women on an average). Though the reasons for reduction in fertility may be manifold, the critical factor is that certain cultural changes have increasingly liberated women from the home and child rearing. According to Tim Dyson, a well-known demographer, in poor traditional patriarchal societies, the spread of TV has opened many women's eyes to a whole new world and through this cultural diffusion, modern birth control methods have allowed them to achieve their aspirations of leading a good life. In fact, with the educated and employed women, 'not to have a child' has become a statement of modernity and emancipation and they are not interested to give up this kind of freedom. Even in rural India, women are running the farms and have taken up the role of their brothers and husbands. This kind of shift may reduce the importance of men and as a consequence, the old Indian preference for sons may diminish. This may help in reducing fertility as couples have started viewing both daughters and sons as potential heads of the new generation (ibid.).

Coale (1973) has mentioned three pre-conditions for fertility decline. First, fertility must be considered within the realm of conscious choice so as to allow individuals to view their life and family away from a fatalistic attitude towards fertility to treating procreation as an object of their life course planning (in most of demographic transitions, some fertility regulation was already practised during the pre-transition phase, more for spacing than for limiting the final number of children). The second pre-condition is that there must be some objective advantages to lower fertility. Third, acceptable means of fertility reductions must be easily available and be at hand.

After having examined the world scenario on population growth and fertility decline, it may be imperative to look into the population dynamics operating in India with a view to highlight the socio-economic consequences for our development and the desirable interventions needed to have an optimal population size for sustainable development. At the time of independence of the country, the population was 342 million. This number has multiplied more than three-fourth during the last six decades or so. According to the Provisional Population Totals (India Census 2011), our population is now nearly 1,210 million (121 crore). The male population is around 6,237 million, while that of females is 5,864 millions. The sex ratio is 946 females per 1,000 males, while the density of population is reported to be 382 persons per square kilometre. The decadal population growth for the period 2001 to 2011 is reported to be 17.6 per cent for the total population, whereas it is 17.19 for males and 18.12 persons for females. It is also reported that the decadal growth rate of India's population has had the sharpest decline during 2001 to 2011 since Independence. (It declined from 23.87 per cent in 1981–91 to 21.54 per cent in 1991–2001 and to 17.64 per cent between 2001–2011). It is also known that the decadal growth of the six most populous states in India such as Bihar, Andhra Pradesh, Madhya Pradesh, Uttar Pradesh, Maharashtra and West Bengal have declined during the decade 2001 to 2011 as compared to 1991–2001.

Interestingly, the net increase in population during 2001 to 2011 has been 181 million people against 182 million in the previous decade, which is equivalent to the addition of one Brazil and one Australia every year.

The six 'Bimaru' states like UP, MP, Chhattisgarh, Bihar, Jharkhand and Rajasthan had nearly 45 per cent of the population of the national total (Bihar leading the rest with the decadal growth of 26%). On the other hand, Kerala, which is very high on social and health indicators, had the least population growth of 4.3 per cent during the decade. One does not understand why the other states cannot follow the Kerala model. This population increase has been observed mostly in poor families. However, West Bengal, which also had a Left government for long, differs in poverty levels and the response to health and family planning initiatives than Kerala. The National Population Policy has set the objective of stabilizing the population level at 145 crores by 2045. But this appears to be an impossible task. Even by most conservative estimates, Indian population may be 136 crores in 2021 and 147 crores in 2031. An increase in population by 18 million per year is indicative of a kind of population explosion in the country.

When the figures for the population increase for the 2001–2011 decade were announced, there was hardly any hue and cry heard from the government, the Planning Commission, leaders of opposition parties, state Chief Ministers, press, TV channels and social activists as if it was not an alarming situation for the country and perhaps the problem belonged to a neighbouring country. The direct outcome of population explosion in our country is going to be great as increasing poverty, hunger, squalor, unemployment, overcrowding, insanitation, social tension and terrorism may pose threatening challenges to the peace and security of the country (Malik, 2011).

We shall highlight some of the significant socio-economic and cultural consequences of population growth in the context of the development of the Indian society.

If we examine the changes in the population structure in India during the last six decades, we observe that as a result of improvement in health conditions due to provision of better medical facilities and control of disease, we have succeeded in bringing down both the birth as well as the death rates. However, due to a fall or check on the infant and maternal mortality rates, the gap between the birth and death rates has widened. But, as a consequence of the birth rate still being very large, there is an increase in population size. (There is also an increase in out-migration from rural to urban areas). This has caused a decline in resource availability per capita in rural areas. Besides, shrinking economic opportunities in rural areas and better health and education facilities in urban areas for higher level of human capital development are the main factors for rural out-migration. But, as most of our population lives in rural areas and is engaged in agriculture, the population rise is observed to be on the higher side in rural than in urban areas.

The debatable point in most of the fora is whether like Europe and America, where contraceptive technology got evolved and was made easily available and where due to economic development there was a greater chance of combining a career with motherhood, should India go in for adoption of family limitation practices or focus on developmental inputs in primary (agriculture), secondary (industrial) and tertiary (services) sectors of economic development?

In view of limited money and material resources at our disposal, our economic growth by and large remained sluggish in the early years of development (the GDP till 1970s was hovering around 3.5%). In fact, during 1970s, food production growth was on an average almost 3 per cent a year as compared to the population growth of 2.2 per cent a year. No doubt, the vagaries of weather are still important in agricultural development but with the spread of irrigation and modern farming techniques, we could raise production in the wheat-growing areas of Punjab, Haryana and western UP. But, of late, especially between 1997–98 and 2002–2003, agricultural growth is observed to be on the negative side. (This is mainly attributable to a saturation point attained in agricultural growth and expansion of secondary and tertiary sectors of economy. High cost of farm inputs and running of tubewell and pumping sets, soil degradation, waterlogging, excessive use of fertilizers and pesticides, non-availability of farm credit and fragmentation of land holdings, etc., are the other reasons).

As a consequence of decline in agricultural output, the real income of farmers, especially those belonging to small and marginal groups, has fallen considerably. This has given rise to suicides among some of them in states like Punjab, Andhra Pradesh, Maharashtra, etc. Further, the plight of the landless labourers has become pathetic as they have become unemployed and this has driven them below the poverty line. As such, agriculture does not hold much promise for development unless and until we go in for a second Green Revolution. Our industrial sector of development was emphasized during 1950–1951 to 1962–1965. If we compare the proportion of contribution of different sectors of economy towards our GDP, we observe that the contribution of agriculture towards our GDP declined from 31 per cent to 22 per cent from 1992–2003. But in respect of industry, for the same period, it remained more or less the same, that is, the decline was from 28 per cent to 27 per cent (for manufacturing, the decline was from 17% to 16%). However, in respect of services, that is tertiary sector, there was a sharp rise in this sector from 41 per cent to 51 per cent as far as its contribution towards GDP is concerned (World Bank, 2005).

It is apparent from the above figures that in the case of the Indian economy, there is expansion of the tertiary sector while the secondary sector remained more or less constant without any expansion. This may partially explain why India could not industrialize and our GDP remained stagnant around 3.5 per cent for many decades. However, due to liberalized economic measures during 1980s, the growth rate went up to around 5.5 per cent while in the post-economic reforms period during 1990s and beyond, we are experiencing around 8 per cent growth rate. We have yet to catch up with the developed world as well as China in the Asian continent to stabilize our population size and bring it to the replacement level. (For example, in Sweden, there are 1.6 children per couple, Norway 1.8, Britain 1.7 and Finland 1.7, which are closer to the replacement level. Nordic governments are observed to be doing a lot in helping couples juggle family and work. In Sweden, women are part-time workers, crèches are nearly universal and paid parental leave lasts for a year. This is not the case in Italy, where 12 per cent of employed women have part-time jobs and in Eastern Europe, the fertility levels have plunged due to the collapse of Communism and the wrecked state-funded family support services. In Singapore, to cope with the decline in family size, for single people, there is a financial incentive for families, increase in maternity leave and downsizing of the working hours).

In view of limited economic growth and development in our country, the developmental issues require huge resource investment because of rising poverty ratios both in urban and rural areas (between 35 and 40% people are reported to be living below the poverty line), unemployment especially among the youth, who are reported to be about 55 per cent of our population and are below 25 years of age, malnourishment among young children, especially those below six years of age (we have nearly 30.12 per cent children below six years of age, out of which males constitute 13.3 and females 12.93 per cent of the total population as per Census 2011), insanitary living conditions, environmental degradation due to fossil fuel burning, global warming effects, ozone depletion, soil erosion and desertification, etc. Our present economic situation does not warrant us to tackle the population growth problem by treating development as the best contraceptive advocated by population experts at the Cairo Conference on Population, as well as by leading economists like Amartya Sen. As such, we need to go in for planned intervention of adopting family planning methods either by utilizing the existing contraceptive technology or developing an indigenous family limitation technology based on our herbal medicinal plants, etc. (In China, the indigenous contraceptive technology like 'paper pill', developed out of cotton seed, is reported to have been successful in limiting the family size to a large extent, besides coercive methods used to have a one-child family norm).

No doubt, India is being viewed as having a greater potential for economic growth than China (we have more than half of our population which is below 25 years of age, while China has a large majority of population moving out of the productive age group to older age groups). But this young potential population number needs to be converted into quality population through appropriate education and skill development and well-planned health care services so that they could contribute significantly in raising the production and productivity in various sectors of our economy for sustainable development. The inputs of social sector such as education and health care are not only desirable but essential for making our economy vibrant and growing at a faster rate.

Population and Needed Interventions

In view of the slow pace of development in the developing nations, it becomes necessary to supplement economic initiatives with certain kinds of technological interventions to check population growth. In this regard, the innovation of contraceptive technology has proved useful in limiting the family size, not only in developed but also in the developing nations.

Historically speaking, birth control methods have been in practice even in traditional societies. These though have been mostly conventional, which includes abstinence of sexual intercourse or withdrawal methods. It is reported that even upper class Romans, in order to have small families, were observed to be concerned with contraception. It is observed by Keith Thomas that contraception was not a modern intervention of the 19th century but it formed a part of the medical tradition way back to Aristotle's time and the Hippocratic Corpus. Even Egyptians observed the practice of contraception. We have also discussions in Arab literature on intra-vaginal female methods of contraception during the 9th century A.D. Despite sexuality being constrained from the dictates of religion-dominant institutions and the guardians, contraception in one form or the other has been in existence (Dutta, 2011).

It was only during 19th century that there were efforts by free thinkers and political radicals in the West to use the concept of contraception for demystifying nature and promoting social justice for the working classes. (In 1822, Francis Place, an English working woman, tried to explain the working class population in industrial cities that sex should be separated from procreation. Later on, Wilhelm, P.J. Mensinga, a German Professor of Anatomy, developed a rubber vaginal diaphragm which became a symbol of modern contraception. At a later stage, we have the marketing of the anti-ovulant birth control pill in 1960). There was a voice for reproductive autonomy for women and contraception was associated with sexual freedom and choice. (Annie Besant, the socialist feminist and a campaigner for India's independence in late 19th century England, was in favour of contraception and women's rights). The term 'birth control' was coined in the 1914 issue of the 'Women Rebel' (a feminist journal) published in New York by Margaret Sangar. Her campaign like her British counterpart, Mary Stopes, was basically for justice to women and this resulted in the opening of the birth control clinic. Sanger's ideas on contraception were not appreciated by Mahatma Gandhi when they happened to meet for a historic conversation. Gandhi considered birth control as a kind of self-control and for that he emphasized sexual abstinence as he regarded birth control measures as dangerous. But to Sanger, birth control was nothing but a contraceptive method. Mary Stopes, considered Gandhi's resistance to contraception, as a kind of an ill-considered arrogance because he considered sexual joy via contraception. Gandhi disliked the idea of sexual dissatisfaction on the part of women and always made efforts to teach them to say 'No'. Gandhian morality was not well taken up by the middle class Indians and it was conceived as some kind of displeasure out of sex and placed women at the alter of self-sacrifice (ibid.).

In the post-independence era in India, the recommendation of the Health and Development Committee, popularly known as Bhore Committee (1943–1946), showed concern about the steady growth of population. Inter alia, it suggested that in order to ensure a decline in the rate of population growth, we should (i) raise the age at marriage for girls, (ii) improve the standard of living of people and (iii) have internal limitation of families. It was also stated that on economic grounds, contraception is justified both in the interest of the individual and community. Therefore, the state must provide facilities for imparting knowledge for birth control when desired for such reasons, that is, economic. (A Family Planning Committee was established in 1949, which was later redesignated as the Family Planning Association of India. The organization had Lady Dhanavanthi Rama Rao as its first chairperson and was having medical practitioners and social workers as its members. It is now spread all over the country and is doing very useful work in this field).

After independence, we had problems of widespread poverty disease stricken population, millions of illiterate people and limited resources at our disposal. However, it was only in March 1950 that the Planning Commission was established and a number of advisory panels were appointed to the Commission with Prime Minister Jawahar Lal Nehru as its chairman. One such panel was appointed for health and the other for social welfare. The health panel appointed a sub-committee on Population Growth and Family Planning on April 11, 1950, which had visualized the problem of impending population increase. This committee, in its report on April 14, 1951, recognized the need for family planning and suggested specific measures in relation to family limitation, that is, facilities for sterilization and advice on the use of contraceptives. Further, it recommended improvements in population data and systematic studies on population problems. The Family Planning Programme got initiated during the First Five-Year Plan during 1951–1956 in the Ministry of Health, with a budget allocation of ₹ 65 lakh. The programme got a boost from the Second Five-Year Plan (1956–1961), and through all the five-year plans till the recent Eleventh Five-Year Plan with increased budget allocations, while integrating the Family Welfare Programme with reproductive health care of mother and pre-natal, natal and post-natal care of the children. The focus was on nourishment of both the mother and children through various essential nutrients. At the same time, population and health care policies were evolved in 1983 and later on, the Population Policy was formalized in 2000. Through various five-year plans, efforts have been made to link the family and population measures with socio-economic development parameters both in agricultural and industrial sectors. Further, improvements in the service sector were pursued to bring down poverty, raise the literacy and educational level of all, especially women, and by providing social welfare measures such as social security, human rights, etc.

Despite efforts made through the five-year plans to develop Indian society by tackling the problems of food, clothing, shelter and nutrition, a large majority of the people (say around 35–40%) are living below the poverty line. Though there is a decline in the birth rate, especially in the urban areas, due to female literacy, education and employment than in the rural areas, the urban areas observed a high rate of immigration from rural areas, which resulted in their overcrowding. Besides, it led to expansion of slums, lack of drinking water and sanitation facilities, lack of job opportunities and shortage of housing accommodation, especially in large and metropolitan cities. In short, poverty, pollution and ill health became serious problems in the urban centres of the country (Sinha, 1992).

As regards the impact of the Family Planning Programme, various quantifiable, that is demographic (population), and qualitative parameters such as economic, social and attitudinal need to be considered. However, the state machinery (Family Welfare Department) estimates that more than 110 million births have been averted between 1956 and 1990 as a consequence of protection provided by various contraception methods such s sterilization, IUD insertion and the use of other conventional contraceptives, oral pills, etc. But, it is observed that the number of births averted does not depend only on the coverage of contraceptives because they do not exclusively indicate a declining birth rate. India had decadal birth rate decline even in the first half of the 20th century when there was no family planning intervention. This may be due to prevalence of killer diseases during that period. The present decline in birth rate may, however, be attributed to a marginal increase in female literacy and employment opportunities for females, along with raising the minimum age of marriage for girls.

We have seen that despite huge investments in the Family Planning Programme, there has been not much impact on the reduction of birth rate. No doubt, several studies have shown that 'female literacy' and 'age at marriage' seem to have a significant role in lowering the birth rate, yet one cannot always correlate couple protection rate with the crude birth rate. In addition to 'female literacy' and 'age at marriage', better economic conditions have lowered the infant mortality rate, which has also significantly helped in controlling the birth rate (ibid.).

Population control measures should focus on voluntary contraception, family planning education, empowerment of women to have autonomy of their own fertility and reproductive decisions, besides providing economic circumstances where 'full education and autonomy' is possible. Further, the religious patriarchs should be discouraged from making harmful statements which may encourage population growth, oppose contraception and treat sex solely for procreation. Abortion remains a controversial issue of the new medical technology because it presents an ethical dimension. Instead of following religious edicts and violent conflicts, there is a need to discuss this as an ethical debate within the framework of 'individual *conscience*'. Rather, the focus should be on good education, free access to contraception and responsible use of reproductive advice. It is rather unfortunate that nearly 14 million children die in developing countries before they reach their fifth birthday. Further, each year, some half a million mothers die in pregnancy or childbirth and nearly 1 per cent of them are in the Third World. (For example, an African woman is 500 times more likely to die from giving birth than her counterpart in the developed countries. Also, nearly one lakh to two lakh women die each year as a result of illegal abortions. Again, the majority of them belong to the developing countries). It is observed that almost all such women and children die unnecessarily. There is need to provide basic health and sanitation services, including breastfeeding, immunization programmes and simple mixtures of salts to prevent dehydration from diarrhoea. Besides education on health and family planning, clean water and enough food to prevent infant and child deaths need to be emphasized. Further, maternal health care and spacing as well as reducing the number of births would help reduce maternal mortality. These are not very expensive educational measures and should be all available as a part of basic human rights (Fielder and King, 2010).

We have observed that India's family planning programme, which was based on voluntary basis to accept contraception, had witnessed a coercive approach during Emergency period, that is 1975–1977, by the ruthless campaign of 'Nasbandi' launched by Sanjay Gandhi. (It is reported that in the last six months of 1976, 6.5 million people in India were sterilized). However, now in a rapidly globalized India, there is an introduction of hazardous hormone and emergency contraceptives which are fatal for women's health. In fact, women's bodies are being used as testing grounds by multinational companies which are promoting life-threatening contraceptive pills to be used by women of the Third World countries, including India. (As revealed by National Family Health Survey, 2005–06, 49 per cent of Indian women are using modern contraceptives and of these only 3 per cent are on the pill. However, a recent survey indicated that 72 per cent people in India do not use contraceptives with a new partner. This often results in unwanted pregnancies and unsafe abortions). It is argued that without advances made in women's education, health care and incomes, will contraception work? In a patriarchal society, is it possible to have sexual rights as women mostly enjoy a subordinate status and submerged location? In a society which is riddled with poverty, structural inequalities and discrimination where women do not have legal rights, including marriage rights to choose their own partners freely, it is very difficult to say what will help or condemn you. In short, contraception is not just a passport to women's freedom as women identities are divided along class, caste, ethnic and religious lines in different locations. In fact, the expressions of their desired pleasure and romance in the everyday life gets silenced by the very forces and institutions advocating contraception in our country, in the name of democracy and progress (Dutta, 2011).

In the context of patriarchal societies, we cannot ignore men as most of the decisions at the family or household level are taken by them. As such, a change in their attitude is most desirable in case we want to transfer certain roles to women from the traditional social order in a modernizing society. It must be kept in mind that all the complexities of the population can be reduced to the outcome of decisions taken by individual men and women.

In the context of developing nations, a three-pronged approach has been suggested to stabilize family size. First, we need to strengthen the health aspects of population, that is, reduction in maternal, infant and child mortality by providing a sound reproductive health care system. Secondly, family planning and information (communication) services need to be made available to all the families. Thirdly, there is a need to devise effective population, education and communication programmes to instill in the minds of couples a small family norm which should be in congruence with the religious and cultural values of people while respecting the doctrine of voluntary decisions (Salas, 1983).

We need political conviction and commitment to work towards these options for attaining the stabilization of population in a democratic country like ours with a view to alter values at the individual or family level, to provide minimum decent living by arriving at a decision mutually between couples to limit the family size.

Population, Poverty and Sustainable Development

Population increase, especially in the context of developing nations, has given rise to poverty syndrome. (According to United Nations Population Estimates, nearly 150 persons are added in a minute to the number of those living on this planet. It has been observed that the world's seventh billionth person was born on October 31, 2011). This is alarming. It presents a gloomy picture so far as ecological predicament is concerned because more people (at the rate of an extra two every second) would mean more mouths to feed, more environmental degradation, more species extinction, more global warming and even a bigger demand on the depleting resources. It is realized that the world population is a very significant factor as it affects both poverty and hunger as well as the habitat destruction and loss of biodiversity. However, the population problem gets more complicated due to severe inequities in economic and energy consumptions across the world. Though population growth has declined in the developed world as a result of demographic transition, which resulted in higher living standard and better education, the developing world is in the grip of various cycles of exploitation which cause poverty, hunger, lack of education, population growth and habitat destruction. As such, the population problem cannot be tackled without focusing on educational, gender and economic equity between the developed and the developing world.

In developing countries, children are considered as economic assets to assist the family in some useful work, from the age of six or seven. By the time they reach the age of ten or twelve, they start producing more for the family than they consume. Further, children provide security in old age and because of high infant mortality, parents need to have a lot of babies to ensure that enough number survives. Having many children may be considered of some sense to individual families, but it is not desirable for the societies in which they live as rapid growing populations often hamper economic development. Many of the developing nations cannot match the resources for the increasing numbers. As a consequence of that, there is mounting poverty, unemployment, slums and scattered settlements, lack of access to education, health care, drinking water and sanitation and family planning services. As poverty deepens, more number of people are pushed to the survival edge. It is observed that even if people want to have children, the best contraceptive may not be of any avail. Rapid population growth is linked to poverty and there is a need to tackle it along with education of women. Research evidence suggests that population control programmes that have ignored social factors have hardly succeeded. For example, nations such as Myanmar, Columbia, China, Sri Lanka, Chile and Cuba and the Indian state of Kerala, which have addressed the poverty syndrome, were able to cut fertility by a third to half between 1960 and 1985 (Fielder and King, 2010).

It is observed that on an average, people in the richest developed nations consume between 30 and 40 per cent more calories than they need, whereas those of the poorest nations, on an average, get 10 per cent less than their basic minimum requirement. More than one billion people, that is about one in every six on earth, do not get enough food to lead fully productive lives. Further, at least 400 million of them get less than 80 per cent of their basic needs and are vulnerable to stunted growth and constant danger of serious illness. (Of these, nearly 2/3rd live in Asia and another one-fifth in Africa. Moreover, nearly two-third are, perhaps, under 15 years of age and nearly about 11 million children under the age of five every year, die from hunger or hunger-related diseases). One study done on malnourished Indian children (which was longitudinal in nature and had followed them for the next seventeen years of their life) found that their capacity for work was 30 per cent less than that of children from the same class and the same villages who have had enough to eat. Paradoxically, people go hungry in a world that produces more than enough because the poor cannot buy or grow the food they need. It is observed that the food production increase by itself may not tackle the hunger problem. We have also to consider, along with a food production, a rise in the consumption pattern. (For example, it is reported that India was a food production success story due to the Green Revolution in 1965–1972, but it turned out to be a consumption disaster). On the one hand, it provided food aid to the newly emergent Bangladesh. In fact, it became the world's second biggest donor after the United States because by mid-1980s, it had gained a surplus of 24 billion metric tonnes but on the other hand, it has still about half of all the hungry people on this planet. It was mainly because the consumption of grain per head failed to increase over the period of production 'miracle' and nearly half of its people are so poor that they cannot buy enough to eat (ibid.).

The international community since the late 1980s has made the eradication of poverty its foremost development objective. About a decade back, the development assistance committee of the Organization for Economic Cooperation and Development (OECD) set the objective of reducing the incidence of income poverty in developing countries from 30 per cent to 15 per cent between 1990 and 2015. There is a need to tackle rapid poverty by focusing on pro-poor growth strategies along with alternative structural, macro-economic, distributive and external policies in a revised development approach. It is much more desirable that in order to tackle the problem of inequality in poor agrarian economies, we must stress on the traditional measures of asset and income redistribution, especially through well-defined land reforms (Cornia, 2002).

The rich developed nations of the world must check their consumption levels so that the poor in the developing nations could use some of the resources to raise their level of living and meet their food requirements adequately. The world is still capable of producing enough to feed seven billion people and more, by increasing agriculture productivity by two-thirds when population peaks at 9 billion by 2050. (We have witnessed food production by five times that amount in four decades up to 2010). Further, it is observed that the number of children expected of a woman is far more important than the crude numbers. It is experienced that when the fertility rate rises, a generational wave surges through the society as was observed in the case of United Kingdom in 1945. When that generation becomes old to work, then you have more economically active adults, smaller families, rising incomes, better standard of living, greater life expectancy, economic growth, increased demand and social change. Economists call it a 'demographic dividend', which we observed in post-war Europe after 1945 and in East Asia in the post 1980s. The emergence of America in the 1800s may be attributed to this rather than to an increase in global free trade. In China, this rising population became sophisticated manufacturers. In India, they developed software outsourcing (Valley, 2011).

In the context of poor countries, individuals need choice and not control. In a situation where life is uncertain, people may like to go in for big families because they have so many jobs to do. For example, fetching water from distant wells, collecting firewood and tending herds. Perhaps, children are the only life insurance available in old age. When almost half your children die before the age of five, it makes sense to have more of them. As such, big families are a symptom of poverty and not that of a cause. Historical evidence suggests a sustained fall in birth rate by a significant fall in child death rate. Women in the poor communities have to be provided full health services and not just family planning. It is necessary that they get rights to food, water, justice and fair wages, along with reproductive rights. Women's literacy and education may help in reducing infant mortalities and birth rates. This is the story of Kerala as compared to the more impoverished northern states in India highlighted by Amartya Sen. Perhaps, development is the best contraceptive and there is a need to focus that in our context for tackling the population dilemma (ibid.).

As per UNDP report 1997, poverty during the last 50 years has fallen more than in the previous 500 years (the reduction is observed in almost all the countries), but absolute poverty is going to confront humanity at the beginning of the 21st century (we have nearly 1/3rd of the world population, that is 1.3 billion people, living on incomes of less than 1 US dollar a day). The disparities between the rich and poor have become more pronounced in the recent years. There is concentration of wealth in fewer hands as reported by UNDP (1998). In 1960, 20 per cent of the world population living in richer countries had 30 times the income of the poorest 20 per cent. However, in 1995, it went up to 82 times. The UNDP report has also brought out that the world's 225 richest people have a combined wealth of over one trillion, equal to the annual income of the poorest 47 per cent of the world's income (2.5 billion).

The increasing concentration of wealth in a few hands and the growing income disparities are not only the problem of the developing nations, but these have also been observed for the developed nations. The poor and poverty in many societies have been explained by the economists but in our context, poverty is also associated with certain socio-religious groups which advocate ritualistic conformity, practice of endogamy and social distancing on polluting dimensions. However, in economic terms, lack of purchasing power on the part of individuals or household families to buy food enough for their survival is the main criterion for differentiating the poor from the non-poor. (There are two standard measures employed to count poverty in a given society. One pertains to the per capita income of the individual or household families, which may vary from time to time due to variations observed in prices of food stuffs. The other measure is based on the minimum calorie requirement for an average adult person, which has been taken as 2,250 calories of food providing various nutrients).

The basic problem of tackling poverty is equitable distribution of foodstuff and measures to acquire these basic ingredients so that the poor can lead an active and healthy life. In the developing nations, poverty has been observed to be perpetuated from one generation to another. However, this appears to be in contradiction to what is desirable in sustainable development of a society because the 'culture of poverty' (a term coined by Oscar Lewis in his study of 'The Children of Sanchez') gets perpetuated from one generation to another along family lines (this concept encompasses a number of associated traits related to family structure, inter-personal relations, time-orientation value systems and expenditure patterns which transcend national boundaries).

In the Indian context, we must consider a poor individual as a person who is not only deprived of his or her purchasing power to buy food enough for sustenance, but also one who is alienated socio-culturally and psychologically to the extent that he or she is discriminated within own community or nation. On the other side, poverty has to be conceived as a phenomenon which isolates certain persons in a population so as to deprive them of their status or position in the society because of their incapacity to produce adequately and their inability to have access to resources and facilities. This further deprives them of a sense of participation, sharing of cultural values and beliefs with others and the freedom to participate in development initiatives (Mehta, 1984).

It is reported by some scholars that poverty at the time of independence of the country was around 60 per cent. Further, it was observed that during 1950–1955, the poverty count was on an average over 50 per cent, which continued till early 1970s (range varying from below 45% to 60%). The decline in poverty count started from that period till the mid 1980s. However, the absolute number in poverty was observed to rise due to increase in population. Based on the 30 days' recall period on food expenditure data collected from the respondents in respect of their food intake, the National Sample Survey for the years 1999–2000 has estimated poverty at 26.1 per cent in our country (Government of India, Economic Survey, 2001). There are, however, doubts on the figure in certain quarters while others consider it as correct. The Tendulkar Committee report prescribed another methodology which has indicated a poverty count of nearly 35 per cent. But the latest controversy was thrown up by Planning Commission Deputy Chairman Montek Singh Ahluwalia and Rural Development Minister Jairam Ramesh. They considered persons consuming items worth more than ₹ 32 a day in urban areas and ₹ 26 in rural areas as not to be poor. But in view of the recent food inflation the purchasing power of the masses does not support this line of thinking. India is ranked 67th out of 88 countries in the Global Hunger Index and nearly a half of India's children remain undernourished, twice the number in Sub-Saharan Africa. Activist Aruna Roy has pointed out that the Planning Commission has itself admitted that households at the current poverty line are getting 20 per cent less food than they require as per the government's own norms. Despite the claims on high growth rate, we cannot lose sight of serious income disparities and low performance per capita on social and economic indicators (Editorial, *The Tribune,* October 2011).

In view of such an alarming population rise and increasing poverty dimensions, is it possible to have sustainable development in the Indian context? We have to conceive sustainable development in terms of dimensions such as 'for whom' and 'how long' and 'where'. Sustainable development also has a time dimension and it should continue to the next generation of people till they assume the responsibility.

Industrialization and urbanization have caused environmental degradation, air and water pollution, acid rain, green house effects due to emission of gases and ozone depletion, etc. We may have limited coal and oil resources, but we have plenty of sun and air; renewable energy systems shall have to be given priority and exploited to the maximum extent. Further, means of communication, especially road network, larger irrigation outlay, dry land farming, soil conservation, agricultural development especially in eastern regions, electrification of railways, improvement in water ways, check on water and air pollution, treatment of industrial effluents for re-fuel, proper sewerage and garbage disposal and that of industrial waste, land reforms, diversification of agriculture and changes in cropping patterns, management of flood, drought and deforestation will go a long way in making India's development sustainable. All these activities will require technology transfer as well as building up technology capacity and manpower indigenously for sustaining these activities. We shall also have to address various other social concerns such as literacy, child survival and nutritional status of pre-school children and women development. Besides, we need to bring in reduction in birth rate and provide minimum health care to improve the quality and productivity of the population for sustainable development (Mehta, 1997).

In the light of the above and changes observed in the contemporary Indian society, sustainable development may not be an impossible proposition but a difficult one. We need to show skills and become efficient managers and the lead in this regard has to be provided by all, including politicians, bureaucrats, leaders of NGOs, intellectuals, etc.

Population and Climate Change: Environmental Challenges and Policy Interventions

Climate change and population growth will show global implication for many generations to come as there is an international agreement on population and on the global environmental issues. However, despite the relationship between the two being recognized, there is not any specific recommendation based on analysis of the linkages which is of any significance. On the environment side, Agenda 21 signed at the Earth Summit in 1992 provided a blueprint for sustainable development, yet the only recommendation made was that nations should take demographic factors into account in the policy-making process. On the population side, the action programme arrived at the International Conference on Population and Development (ICPD) in Cairo, 1994 highlighted population and environment links but it was nothing more than a repeat of the language of Agenda 21 (Lutz, 2002).

It is well-known that human beings exploit natural resources through the consumption of goods and services. As a result of that, pollution including emissions of carbon dioxide and other green house gases, which result from the production of goods and services as well as through their consumption, may give rise to various kinds of residuals. Production of goods and services ultimately depends upon consumer demand as well as on government and military expenditure. That is why the role of consumption, which is referred to as 'consumer sovereignty' has been observed to be very significant in the neo-classical economic model of environment. It permits economists to suggest that consumers can make intelligent trade-offs between consumption of goods and the quality of the environment. Production is not only driven by the kinds of choices made by the consumers, but is also determined by the scale of consumer demand which forms part of the population size. For example, if fertility is to be conceived as a human choice, then there is a kind of inverse relationship between the number of people and the quality of the environment and between the number of people and the material standard of living. This is how neo-classical population – environment-consumption trade-off – gets formulated. There is both a direct and indirect effect of population on the environment. The direct effect was the prime concern of Malthus and is mainly expressed by ecologists and ecological economists, whereas there are a lot of ambiguities and disputes regarding the nature of indirect effect and it is this that requires the focus of attention (ibid.).

Apart from these direct and indirect effects, there are induced effects which focus on the changes in the way social institutions cope with ecological stress, including changes in the institutions themselves, as a consequence of shifts in demographic regimes. In certain circumstances, it has been observed that rapid population growth results in social and institutional changes that may promote environmentally sound adaptation, while in certain other cases, it gives rise to institutional and social situations, ensuing environmental degradation. With the slowdown of population growth, there is likelihood of a reduction in green house gases in the long run. Further, it improves the resiliency of vulnerable populations to climate-related impacts. As such, climate change strengthens the case for voluntary family planning programmes by increasing the returns to investments. Besides, it strengthens the case for indirect population policies, especially investments in female education, as they are not only going to increase individual welfare but may also result in lowering fertility. Climate change through a 'green house' externality to child-bearing is also going to add to the cost involved and benefits obtained in the case of fertility.

This is a ticklish area because economic costs and benefits of child-rearing cannot be estimated in a comprehensive manner. Moreover, there is innate reluctance to view fertility decisions as purely economic in nature. As such, the externality studies may be useful for throwing light on the consequences of child-bearing, but they cannot be considered as the main basis for policy decisions. However, the studies through 'green house' gas emission lend a quantitative support that population growth imposes external environmental costs on society though largely it remains an unmeasured environmental externality. However, this strengthens the case for programmes that result in lower fertility. Further, the threat of climate change strengthens the equity-based rationale for population policies because its consequences are likely to fall most heavily on the least well-off persons. In the current dominant thinking in the population policy, the demographic consequences are given hardly any weight whereas these are the ones that provide the main link to the climate change issue. No doubt, the empowerment of women and providing reproductive rights and health care to them were the main justifications for programmes called for by the Cairo conference, yet, from the point of view of climate change, the kind of population policies most favoured had a win-win situation. It is observed that providing reproductive health services (including family planning, investing in education, especially for women), reducing infant and child mortality and equalizing gender relations may result in lowering fertility. As such, these policies not only increase the welfare of individuals directly, but may also reduce future environmental stress through lower fertility (ibid.).

It is also known that more than a third of the planet's land surface, including its most productive land, is now dominated by human monoculture. It is expected that in certain parts of the world such as Sub-Saharan Africa, the total fertility rates remain as high as 5.4. Even in the Middle East, the fertility rates at 3.5, are very high. These high rates reduce economic growth and stress environmental resources. Further, the young populations having access to becoming adolescent men may cause manifest political instability and violence. Despite the fact that the world rate of population increase has come down from 2.1 per cent per annum to 1.3 per cent, there is an overall increase in young populations having high birth rates (Fielder and King, 2010).

Though there is a slowing down of world population growth as a result of social factors such as media, increasing education and role of women in society, there is a population explosion and its consequences have resulted in having an inevitable human impact on all aspects of the biosphere. In fact, there is a serious crisis observed as a consequence of population rise on this planet and without tackling it, we cannot save the environment.

In the context of rising population growth in India, which is the second most populous country in the world after China, we have observed that the pangs of poverty and rapid population growth have contributed to environmental degradation through overexploitation of natural resources. There is also a change in the consumption pattern of the people. One such indicator may be the number of motor vehicles registered with the government. This may show a rise in our expanding consumption and economic growth. It is observed that the increasing number of vehicles has produced more air pollution, fuel consumption, traffic jams and demands for road construction, often at the cost of agricultural land. As per Transport Research Wing, India (1997), there is an increase in the total number of registered vehicles from 3 million in 1950–51 to more than 33 million in 1995–96, out of which 28 per cent are concentrated in the 23 metropolitan cities. Further, according to Energy Information Administration 2001, there were about 10 motor vehicles for every 1,000 people in 2000 having a population of just over one billion, whereas in the population of India, which is likely to be 1.3 billion by 2020, there may be 44 motor vehicles for every 1,000 people). This kind of increase will cause environmental problems like air pollution and global warming. Besides, there will be the degrading effects of fuel like oil and petroleum on the environment of the country. Further, rise in demand for energy through fuels such as coal, lignite, oil and nuclear energy, etc., may cause particulate and gaseous pollution, leading to bronchitis and respiratory diseases. The increasing population number and growing affluence have already resulted in rapid growth of energy production and consumption growth in India and may accelerate in the future. We need to provide some alternative energy resources to cut down energy generation from burning of fossil fuels, especially coal, emitting lots of carbon dioxide, which may result in global warming (in the Indian context, the share of thermal power has increased from 51 per cent in 1950–51 to about 70 per cent in 1995–96, whereas the share of hydropower declined from 49 per cent in 1950–51 to 19 per cent in 1995–96. The share of nuclear power is nominal) (Nagdeve, 2007).

In the wake of the above environmental effects observed, there are a number of environmental challenges such as pressure on land, habitat destruction and loss of biodiversity, water scarcity and climate change that need to be addressed in our context.

India faces the most acute pressure on agricultural land. It is observed that while the total population increased by around three times, the total area of land under cultivation increased by only 20.27 per cent from 118.75 to 142.82 million hectares during the last 15 years, out of the total geographical area of 329 million hectares. There has been agricultural intensification and extensification through increase in cropping and irrigation intensity and higher use of chemical fertilizers, pesticides and insecticides. This has led to land degradation, overexploitation of underground water resources, and water pollution. Besides, there has been waterlogging, salinification, alkalization of crop lands, entrophication of water bodies and marine pollution. This all has resulted in having reductions in biodiversities.

Deforestation has been another problem encountered in our setting. Forests are not only a source of providing raw materials to industries, but these also generate income and employment. Besides, we need to provide an ecological balance of life and life support system by checking soil erosion, maintaining soil fertility, conserving water, regulating water cycles and floods, balancing carbon dioxide and oxygen content in atmosphere. In fact, forests have been a major source of food, fodder, timber, fuel, dyes and oils for medicine and their cutting can result in robbing poor people of their livelihood and the source of medicinal plants. (India has a forest cover of 76.52 million hectares but over the years from 1993 to 1999, there has been a decrease in the forest cover by 2,814 square kilometres such as in states like Andhra Pradesh, Bihar, Kerala, Madhya Pradesh, Odisha and Tamil Nadu. On the other hand, states like Gujarat, Haryana, Himachal Pradesh, Maharashtra and Rajasthan have shown an increase in forest cover for the same period). In fact, deforestation has resulted in causing frequent floods, soil erosion, heavy siltation of dams and changes in climatic conditions, thus resulting in damage to the economy (ibid.).

It is also observed that as a result of rapid population growth, food production cannot match with the demand, as more and more of marginal land is put to cultivation to feed more people. Overexploitation of agriculture land is causing environmental degradation, rising from intense farming activities contributing to soil erosion, land salination and loss of nutrients. (Out of the total geographical area, 175 million hectares is considered to be land-degraded area). Further, there is loss of habitat destruction and biodiversity. Biodiversity is considered to have direct consummative value in food, agriculture, medicine, industry, besides having aesthetic and recreational value. The destruction of plants and animals may not be that threatening to biodiversity, but it is the destruction of their habitat that poses a serious question. (India is considered as one of the twelve mega biodiversity countries of the world. We have nearly 17 per cent of the total geographical area having 46,000 plant species and 81,000 animal species, which represent about 7 per cent of the world's flora and 6.5 per cent of the world's fauna, respectively). Population growth over the years has led to expansion of human settlements and increasing demand for food, fuel and building material. By and large, the habitats have caused damage to the biodiversity due to developmental programmes such as building of reservoirs, mining operations, forest clearing and communication and transport networks, etc.

Again, it is reported that water use in India has increased over the past fifty years. The largest share out of the total annual fresh water withdrawal goes to agriculture, which stands at 92 per cent whereas the industrial use is only up to 3 per cent, while the domestic use count is 5 per cent. The amount of water available per person has declined in recent decades mainly because of population growth and the water scarcity is likely to worsen in the future.

Water pollution is also a serious problem which emerges mainly from domestic sewage, industrial effluents and activities such as agriculture. Besides, seepage is caused by fertilizer plants, refineries, pulp and paper mills, leather tanneries, metal plating and other chemical industries. Because of widespread pollution, access to safe drinking water is only for 70.1 per cent of the households in urban areas and 18.7 per cent in rural areas through organized piped water supply, while others are dependent on surface and ground water which is untreated. (This is as per Central Statistical Organization, India, 1999). As a consequence of that, there is a threat to public health and diseases commonly caused due to polluted water like cholera, diarrhoea, hepatitis, typhoid, amoebic and bacillary dysentery and guinea worms. Further, due to scarcity of water, there is prevalence of scabies, leprosy, trachoma and conjunctivitis. By and large, the infant mortality deaths and intestinal diseases are attributed to polluted water (Nagdeve, 2007).

Apart from water pollution, air pollution is also a serious problem in India. Indian cities are reported to be among the most polluted in the world. Air in the metropolitan cities is highly polluted and a pollutant concentration exceeds the limit considered safe by the WHO. The most important air pollutants are residual suspended particulate matter, suspended particulate matter, nitrogen oxide, carbon monoxide, lead and sulphur dioxide, etc. These are mainly caused by growing industrialization, increasing vehicular pollution, industrial emissions, automobile exhaust and the burning of fossil fuels. These cause death to many and respiratory damage, heart and lung diseases to others. It is reported that in Mumbai and Delhi, nearly one-half of the children under the age of three show signs of harmful exposure through lead. Besides, it is also observed that the indoor air pollution caused by cooking and heating with wood, crop residues, animal dung and low quality coal produce smoke which may pose an even more dangerous hazard for human health. (According to a World Bank study, the environmental damage in the year 1992 in our context is estimated at around 10 million US dollars, which is 4.5 per cent of the GDP. Out of this total environmental damage, urban air pollution costs India 1.3 billion US dollar a year, water degradation 5.7 million US dollar, soil erosion 83 to 163 million hectares of land every year and a productivity loss of 2.4 million US dollars or 4 to 6.3 per cent of the productivity every year (UNDP, 1998, as quoted by Nagdeve, 2007).

A brief account of the environmental challenges outlined above as a consequence of population growth and rapid increase of energy use is believed to be causing global warming, having both positive and negative physical, environmental and socio-economic consequences. It is very difficult to have correct estimates of the effects of these due to many marked uncertainties. Further, climate change is also likely to occur that may result in affecting changes in precipitation patterns, ocean circulations and marine systems, soil moisture, water availability and a rise in the sea level. These changes can influence agriculture and nature-ecosystems like wetlands and fisheries even in our environmental setting. In view of these alarming environmental situations, there is a need to formulate policies by the state apparatus to protect natural resources and environment by showing concern to environmental areas, such as agricultural development, urban planning, technological innovations, industrial growth and resource management.

Besides, we must control population growth below the replacement level in the country.

In a recent United Nations climate meet in Durban (December 11, 2011), a compromise deal was struck and a roadmap was prepared as an accord which for the first time legally may force all major carbon emitters to cut green house gas emissions. This brought an end to wrangling between India and the European Union over the fate of the Kyoto Protocol (Japan). This new pact will be finalized till 2015 and for the first time, it will bring India and China under the ambit of a legal mechanism which will guide emission cuts, while the accord will come into effect only from 2020 onwards.

This compromise advocates that while India and China have agreed to be a part of the governing treaty, the developed nations have also shown a second commitment under Kyoto Protocol, thus putting to rest doubts that they would walk away from commitments once the 1997 treaty expires in 2012. The Kyoto Protocol had set the binding targets for cutting green house gas emissions to 5 per cent below the 1990 levels by 2012 for the 37 industrialized nations and the European Union.

However, Sunita Narain, an Indian environmentalist, in her special reporting to *The Tribune* ('Global Distrust Peaked in Durban', December 12, 2011) is of the opinion that the final agreement is of a mixed nature. Although the opposition of India and other countries to a legally binding agreement has been accepted without adequate safeguard as a matter of 'principle of differentiation', yet it has also been highlighted that the future outcome could be a range of options from a process to develop a protocol or another legal instrument or as agreed outcome backed by legal force. The challenge before us is that these future agreements take into account the historical and current emissions of countries in setting targets. Another challenge would be to set quantification limits of the emission and reduction targets for all the industrialized nations so that they can reduce emissions between 25 and 40 per cent over 1990 levels by 2020. The setting up of Green Climate Fund is another outcome of this conference to mitigate and adapt in the vulnerable countries of the world. However, this fund is without any money as the industrialized nations have declined to pay because of recession. This raises another question as to how there can be innovative sources of financing. Narain, in her critical note, has remarked that if the future agreement, which is likely to be negotiated in the next two years, is without a condition of equity in past and future carbon budgets, then the city of Durban may become famous for starting a new era of 'climate apartheid' (Narain, 2011).

The biggest challenge before India is to bring down the population size to a zero replacement level. As the rising population cuts across all development plans in different sectors of economy (primary, secondary and tertiary), this appears to be major source of environmental crisis. Over the years, the state apparatus and the political masters and administrators have appeared to be marginalizing the population control measures and emphasizing more on developmental inputs for the growth of our economy. However, the economic crunch which we are confronting – especially in the social sector such as education, health and population control, so necessary for the development of our society – would warrant us to strengthen our family planning services, educate people through personal and impersonal channels of communication and provide integrated health care to infants, mothers and children for ensuring better survival ratio. Can we move towards that end? For that to happen, perhaps we need a strong political will and a transparent and accountable bureaucratic apparatus to implement the population policies and programme to bring down population growth to a replacement level. If we succeed in that direction, the environmental and climate change issues will automatically get solved or reduced to a low level so as to provide minimum decent living to the majority of our population.

References

Central Statistical Organization, India (1999), *Environment Statistics in Statistical Abstract of India*, New Delhi.

Coale, A.J. (1973), 'The Demographic Transition', in Proceedings of the International Population Conference 1, Liege: International Union for the Scientific Study of Population.

Cornia, Andrea Giovanni (2002), 'Inequality, Growth and Poverty in an Era of Liberalization and Globalization', in *Human Development and the Environment: Challenges for the United Nations in the New Millennium*, Hans Van Ginkel, Breudan Barrett, Julius Court and Jerry Velasquez (Eds), Tokyo: United Nations University Press (Indian Edition, Jaipur: Rawat Publications), pp. 76–109.

Dugan, Emily and David Randall (2011), 'A Billion Go Hungry on World Food Day', *The Tribune*, Chandigarh, 17 October, p. 13.

Dutta, Nonika (2011), 'The Pill that Failed', *The Tribune*, Chandigarh, 14 December, p. 11 (OPED Women).

Fielder, Christine and Chris King (2010), 'The Rape of the Planet and Genetic Holocaust', in *Sexual Paradox: Complementarity, Reproductive Conflict and Human Emergence*, available online at: dhushara.com/book/ diversit/bomb.htm

Government of India (2001), Economic Survey (2001–2002), New Delhi, Ministry of Finance and Company Affairs.

Harvey, Brian and John D. Hallett (1977), *Environment and Society: An Introductory Analysis* (Population and Food), London: The Macmillan Press Ltd., pp. 17–36.

Kang, Manjit S. (2011), 'Complicating Food Crisis: Time for Devising Long-term Policy', *The Tribune*, Chandigarh, 3 October, p. 8.

Lutz, Wolfgang (2002), 'Population: A Global Challenge for the Twenty-first Century', in *Human Development and the Environment: Challenges for the United Nations in the New Millennium*, Hans Van Ginkel, Breudan Barrett, Julius Court and Jerry Velasquez (Eds), Tokyo: United Nations University Press (Indian Edition, Jaipur: Rawat Publications), pp. 126–148.

Malik, Ram Nivas (2011), 'Population Still a Ticking Time Bomb', *The Tribune*, Chandigarh, 18 November, p. 11 (OPED Development).

Mehta, S.R. (1984), *Rural Development Policies and Programmes: A Sociological Perspective*, New Delhi: Sage Publications.

Mehta, S.R. (Ed.) (1997), *Poverty, Population and Sustainable Development: Essays in Honour of Professor V.S. D'Souza*, Jaipur: Rawat Publications.

Nagdeve, Dewaram A. (2007), 'Human Impacts on the Environment in India', in *Environment and Sustainable Development*, Vol. III, K.A. Rasure (Ed.), New Delhi: Serials Publications.

Narain, Sunita (2011), 'Global Distrust Peaked in Durban', *The Tribune*, Chandigarh, December 12, p. 18.

Registrar General and Census Commissioner of India (2011), *Provisional Population Tables*, New Delhi.

Salas, M. Rafael (1983), 'The 1983 State of the World Population', Report, United Nations Fund for Population Activities, Excerpts 'The Family of the Future', *Development Forum*, Vol. XI, No. 6, pp. 8–9.

Sinha, N.K. (1992), 'Family Welfare', in *State of India's Health*, Alok Mukhopadhyay (Ed.), New Delhi: Voluntary Health Association of India, pp. 189–227.

The Tribune (2011), 'Plan Panel Silent: Poverty Will be Redefined', Editorial, Chandigarh, 4 October, p. 8.

Transport Research Wing, India (1997), *Pocket Book on Transport Statistics*, New Delhi: Ministry of Surface Transport.

United Nations Development Programme (1998), 'Unequal Impacts of Environment Damage', *Human Development Report*, New York: Oxford University Press.

Valley, Paul (2011), 'Misplaced Scare over Rising Population', *The Tribune*, Chandigarh, 31 October, p. 9.

World Bank (2005), *World Bank Statistics*, Washington DC.

5

Sociology and Environmental Sustainability

A Cultural and Biodiversity Perspective

Environmental degradation, hazards and damage, affect all species on planet Earth but solutions to these have to be provided by humans. Humans have the programming capabilities and combined with the ingenuity of various minds, they can invent and discover ways as well as means to solve or tackle the environmental crises for making human life sustainable and comfortable for a longer period.

It is often debated whether environment is somewhere outside 'in nature', that is untouched or non-affected by humans or humans are part of it. The relationship of environment with sociology is, however, characterized by the fact that sociology provides a way of viewing and understanding the social world. Further, it allows a better understanding of social organization, inequalities and all types of human interactions. Besides, it has a potential for providing a more just world. The underlying assumption for most of the sociologists to understand fully the relationship between environment and society is that humans are part of the environment. In fact, it is argued that lack of understanding about the human and nature relationship has contributed to the worst environmental problems such as climate change, ozone depletion and so on and has also limited our abilities to solve them (McCarthy and King, 2009).

In the above backdrop, we may like to highlight the perspective that may tackle the environmental problems, issues and crisis being confronted by different societies, especially the developing ones. Developing societies, as we know, are starved of both technological inputs as well as finances. Further, clean technology is available only with the developed nations and the cost at which they make this technology available to the developing nations may go beyond their affordable means to tackle some of their environmental hazards or concerns. Moreover, in the context of what has been discussed so far in the preceding pages on environmental sociology with special focus on health risks and population dilemma (concepts, perspectives, methods and strategies, etc.), the question that pinches us most is the type of approach that the developing nations, like ours, should adopt and pursue to solve the environmental problems confronted by them. Should our environmental policy be guided by radical environmentalism where the state assumes the sole responsibility for tackling the environmental problems, or should it be guided by the principle of individualism advocated by liberal democratic governments as a part of environmental philosophy? Of late, civil society movements on environment protection and preservation have emphasized on the radicalization of individuals and collectivity, to lead them for integrating the environmental issues and crisis. In fact, a more pragmatic approach is being suggested. This envisages public discourse on community-based analysis of the environmental issues having a bearing on community social life. Besides, it is based on a set of practical-oriented action solutions which are influenced by the method of experience of socio-cultural learning in a normative manner for seeking the truth. Further, it is observed that within the praxis-oriented tradition of pragmatism, one may not be in a position to separate fact from 'value' but facts gain their meaning only in an action-oriented context (Norton, 2003).

Another question that is of concern to a developing nation like ours, which is already embedded with inequalities and inequities of various kinds, is related to the impact of new technologies such as nanotechnology, biotechnology, informatics and cognitive and neurosciences, for enhancing human performance as a part of futuristic technological trends. The convergence of these technologies may further be divisive for the developing societies, including ours, as only the affluent class may be able to enjoy the benefits of these for enhanced human performance. However, the majority of the people lower in different strata may not be in a position to avail of them for lack of financial ability. The skewed availability of such technologies may further pose serious environmental problems for the majority of the population, who may become susceptible to higher health risks of varied nature. In such an emerging situation, can we provide environmental justice and be more morally responsible in providing a sustainable environmental policy? Should, we, as humans, not be guided by some kind of environmental ethics and principles to make environmental policy more equitable oriented?

Further, as humans, we need to show concern for our future generations. In fact, we should exhibit bequest to the future and raise a question whether we can do something or not in that direction. Should we be guided only by ecosystem and bio-system ideologies for salvaging and tackling environmental problems or should we build up environmental strategy by taking into account the existing pluralism in developing societies, including ours, for environmental sustainability by strengthening both the prevailing cultural and biodiversities? No doubt, the environmental problems are, by and large, technical in nature but these need social solutions. The questions raised above in the context of environmental sustainability need to be addressed within a sociological perspective, which has to be people-directed, people-focused and people-action oriented by shifting the strategy from liberal democracy to participatory democracy, wherein people could be drawn from all strata. We shall take up these questions and make an attempt to discuss them in detail in the succeeding pages.

Radical Environmentalism (Deep Ecology), Individualism, Eco-feminism, Pragmatism as Part of Philosophy of Environment

Radical Environmentalism

This is a new and growing movement that has shown resentment and disenchantment with the established environmental approach, besides providing a critique on its functioning. This movement is also being mentioned as 'new ecologists' as well as 'deep ecology', which decry the narrow economic goals of mainstream environmentalism. It also provides a philosophical and cultural revolution in human attitude towards nature. In fact, it provides a militant defence of 'Mother Earth' and strongly opposes human attacks on undisturbed wilderness. Its goals range from the spiritual to the political. 'Deep ecology' as it is nomenclatured forms a wide spectrum of the American environment movement.

Guha (1996), as a representative of the Third World, provides a critique by suggesting that 'deep ecology' is uniquely American. Further, despite some superficial similarities in rhetorical style, the social and political goals of radical environmentalism in other cultural contexts (for example, in West Germany and India) are quite different. Also, the social consequences of putting 'deep ecology' into practice on the global basis may indeed be very grave.

According to Guha, 'deep ecology' or radical environmentalism is based on four main principles. One, the environmental movement must shift from an 'anthropocentric' (the belief that man and his works are at the centre of the universe) to a 'biocentric' (assigning man a more humble place in the natural world) perspective. The thrust is on preserving nature as it will benefit the future human generations. Secondly, the focus is on the preservation of unspoiled wilderness, that is restoration of degraded areas to a more pristine condition. Thirdly, there is widespread invocation of Eastern spiritual traditions as frontrunners of 'deep ecology'. This was perhaps to consolidate the claim that 'deep ecology' is a philosophy of universal significance. Fourthly, deep ecologists, despite their internal differences, consider themselves as the leaders (spiritual, philosophical and political) of environmental movements of American and world environmentalism (ibid.).

Guha, in his critique of radical environmentalism or 'deep ecology' (as opposed to 'shallow ecology', a strategy of environmental establishment in the American society guided by individualism as a part of democratic liberal ideology) argues that the transition from an anthropocentric (human-centred) to a biocentric (human as only one element in the ecosystem) view in both religious and scientific traditions, is appreciable. It provides a check on man's arrogance and ecological hegemony. However, it is not acceptable that intervention in nature should primarily be guided by the need to preserve biotic integrity rather than by the needs of the humans. For deep ecologists, the biotic integrity is biocentric, while human needs are anthropocentric. This dichotomous distinction is meaningless in understanding the dynamics of environmental degradation.

The world is confronted by two basic ecological problems. First, there is overconsumption being practised by the industrialized world and by urban elites in the Third World. Secondly, there is a growing militarization in the ongoing regional wars and in the arms race and there are chances of nuclear annihilation. Both these problems do not have any visible connection to the anthropocentric-biocentric distinction. In fact, ecologically wasteful characteristics of the industrial society and of militarization, at the aggregate level, are due to dialectic of economic and political structures. But, at the micro level, they are due to lifestyle choices of individuals. Further, his argument is that if the above dichotomy (anthropocentric vis-à-vis biocentric) is irrelevant, then the emphasis on wilderness is positively harmful in the context of the Third World.

The preservationist-utilitarian divide may reflect a conflict between the 'people' and 'interests' in the developed American society, but in a country like India, this may mainly be a reverse situation. India is a long-settled and densely-populated country where agrarian populations have a harmonious balanced relationship with nature and, as such, the setting aside of wilderness areas may result in a direct transfer of resources from the poor to the rich. (For example, Project Tiger, a network of parks highly rated as successful by the international community, brings into focus the interests of the tiger against those of poor peasants living in and around the reserve).

Guha is also not happy at the way there is persistent invocation of Eastern philosophers as an antecedent for the protection and conservation of environment, by deep ecologists. Even, complex and internally differentiated religious traditions such as Hinduism, Buddhism and Taoism have been clubbed together to support a view of nature being biocentric. Further, there is a claim made by deep ecologists that at the level of natural and spiritual practice, 'primal' people subordinate themselves to the integrity of the biotic universe they lived in. This appropriation of Eastern traditions by them is partly influenced by the need to construct a lineage and partly by a desire to present 'deep ecology' as a universalistic philosophy. More important than all this, Guha raised a fundamental question related to the extent of radicality pursued and practised by the deep ecologists. Except for building their self-image, it appears rhetoric as within the American society, their radicalism is limited while elsewhere, its manifestation is quite different. Guha opines that 'deep ecology' perhaps is best viewed within the wilderness preservation movement. Advancing philosophical rather than aesthetic justifications and encouraging political militancy rather than negotiations appear identical in the preservations of unspoiled nature.

In the mainstream movement, the function of wilderness is to provide a temporary respite to modern work-a-day world. As a part of industrialized society, the 'national park' as an institution developed due to economic expansion of the American society so as to allow people to simultaneously enjoy the material benefits of an expanding economy and the aesthetic sensibilities of unspoilt nature. But the roots of the other world's ecological problems lie in the nature disproportionate consumption of resources by the developed industrialized nations, as well as by the urban elites in the Third World or developing countries. Since it is not possible to have an industrial monoculture across the world, the ecological movement in the West should clean itself by cutting down on the current consumption levels.

This radical shift both in the consumption and production patterns need alternate economic and political structures based on smaller scale, social participation and a shift in cultural values. In fact, the modern Western man needs to imbibe an ethic of renunciation and limiting one's needs by inculcating spiritual and communal values, significant in sustaining social life. This kind of revolution in cultural values may provide a better understanding of environmental processes than that of 'deep ecology' (ibid.).

Notwithstanding the observations máde by Guha on radical environmentalism or 'deep ecology', a movement that is spearheaded the American society as against the mainstream or established environmentalism like any other development parameter (economic, social, political), has to be culture-specific to a country especially in the context of developing nations. The moot question is whether or not we can evolve an environmental policy based on our cultural traditions in the era of globalization and modernization when nature is losing its identity by the greedy power of individualism.

Another environmental philosophical question, as a part of 'deep ecology' axiom, is the identification of individuals who are interconnected with others, as they are not discreet substances who are capable of existing independently of the other individuals. Apart from that, one significant pre-supposition followed by deep ecologists is that nature can look after its own interests. It is, by and large, our interventions in the natural course of events that result in terminal ecological disasters. This kind of supposition implies that nature should take the lead in ecological matters and we need to minimize our interference and carve out our own interests to those of nature. In fact, we should follow the law that 'nature knows best'.

In these two assumptions, interconnectedness with the rest of the nature would mean that ultimately, we are identifiable with nature. This indivisibility between individuals and nature should suggest us to shed off our confining ego identity and gradually we should largely open up to nature. This enhanced possible identification with nature is a sort of self-realization which enlarges one's sphere of identification. In the light of the interconnected thesis, if what we do to the environment is natural and since 'nature knows best', our present disposition of the environment must be in nature's long-term interests.

However, if whatever we are doing to the environment is natural and yet we insist that it needs to be changed by us, then we may be denying that 'nature knows what it is doing'. This may subtly indicate a sort of control over nature. This may also directly imply that we have no ecological as well as moral grounds to intervene in the spontaneous course of human affairs as these may affect the environment. If the deep ecologists tend to acquiesce in the current affairs of environmental degradation and destruction, then one is tempted to conclude that there is a kind of an inconsistency in their approach (Mathews, 1996).

The above observation appears to confirm the critique provided earlier that radical environmentalism or 'deep ecology' is best to be viewed as a radical trend within the realm of the wilderness preservation movement. Mathews (1996), in his analysis of 'deep ecology', shows concern of an intractable dilemma existing in the two assumptions discussed before, that is 'interconnectedness of relations within a natural world' and 'nature can best look after its interests'. However, the question posed is whether nature knows best for itself as a whole or does it know best for the individuals that constitute its elements. Further, it is argued that nature, in a holistic sense, 'knows best' means that it is capable of looking after its own interests. Moreover, in a wider moral sense, the ecological order may not only secure its own self-perpetuation but may also appear to exemplify both justice and generosity.

But when nature is viewed from the individualistic sense, then is it still considered a moral order? In a holistic sense, individuals are generously given life and they justify passing on the life as sacrificial. But when we identify ourselves as individuals, we may see things differently because then the nature may not look after the interests of actual vis-à-vis possible individuals. As actual individuals, we are guided by actual interests, urgent needs and desires and we can suffer terribly. We cannot tradeoff for justice or generosity in the case of actual individuals than possible ones. However, we have fellow feelings for them, show familiarity with their impulses and drives and have identification with their shivering vulnerability. This gives rise to a concern and moral interest for them because of their plight. In fact, our impulse to resist the progressive destruction of the present order of life is not due to identification with nature as a whole (as claimed by 'deep ecology') but rather due to our commitment to our individuality, because as individuals we show our allegiance to individuals even against the moral requirements of nature as a whole (ibid.).

In addition to 'deep ecology', another philosophical environmental question having linkage with nature, over the last couple of decades, is that of Eco-feminism. Mathews (1996), while contrasting it with 'deep ecology', conceives its portrayal as a natural world of community of beings. In fact, our mode of relating with them should be based on open-mindedness through attentive encounters. Such encounters may provide the necessary understanding to build up an attitude of care or compassion to develop an ecological ethic.

Implicitly or explicitly, eco-feminism is connected with the interconnectness thesis already highlighted as an axiom for 'deep ecology'. However, the interconnectedness in eco-feminism is interpreted in the individualistic rather than in the holistic sense. Nature here is conceived as a community of beings who are related to one another as individuals do so in a family setting but nevertheless, they remain distinct. It is emphasized that we have to respect the otherness, that is, the individuality of these beings and not seek any merger with them in order to pursue an undifferentiated oneness. Eco-feminism does not have direct identification with nature as a whole. However, it does not suffer from any kind of identification dilemma as it springs out of a sense of solidarity with our fellow beings (ibid.).

This philosophical construct apart, eco-feminism as a movement was born in France in September 1973. This movement was observed to be close to the Belgian Unified Feminist Party than to the French Movement de Liberation de Femmes (MLF). The new group, the Feminist Front, was a formation of several women from MLF groups and others from subgroup Evolution (founded in 1970 after 'The Women's Estate General' and in reaction to it) and by independent women from all parties and all movements. (The group's statutes followed the 1901 Law for Associations and this new movement was considered much more legalist and even reformist). D'Eaubonne

(1996) in her treatise 'The Time for Eco-feminism' rightly sums up at the end of her essay that 'the problem for a woman is first demography then nature and then the world; her urgent problem, the one in common with youth, is autonomy and control of her destiny'. It is emphatically expressed that if humanity is to survive, then she must develop sensibilities towards this fact while the human being shall finally be treated first as a person, and not at all as a male or female. The society has to place itself in the feminine perspective as no other category can accomplish the ecological revolution, as no other was as directly concerned at all levels. And finally, the planet placed in the feminine perspective will flourish for all (ibid.).

This is how eco-feminism has been explained to have a better understanding of women in environmental protection and preservation for sustainable development. In the development process, in the developing nations, especially in our context, where women constitute nearly half of our population, they have the potential of playing an active role in preserving and protecting the environment. They are close to nature and ours being a bio-massed and agricultural economy (as a part of our cultural tradition), they know the state of art in managing waterways, dairy farming, milching of cows and buffaloes, fishery, poultry farming and many other allied agricultural and animal husbandry activities.

In her critical note, Shiva (1996) makes a mention that the earlier assumption that the development process will provide availability of goods and services in an increased manner to remove poverty no longer holds good for the women's ecological movements in the Third World, though it is still accorded primacy in the centres of patriarchal powers. She is of the opinion that survival is based on the assumption of sanctity of life while 'maldevelopment' embarks the sacredness of 'development'. We should abandon sacredness of development because it threatens survival itself. People want to live as they are tired of development (based on intensive processes of production engineered by energy resources), which direct resources away from survival and hence, from women.

'Pragmatism' as an environmental philosophy is another approach that needs to be taken account of. It recast itself away from environmental ethics to a pragmatic philosophy of policy discourses during the late 1980s and early 1990s. Instead of treating the general nature of environmental value, the focus shifted to have improved problem formulation to look for more sustainable specific situations. When we conceive sustainability of environment, we have to have presence of a forward-looking community that shows respect to the norms involved in the pursuit as well as to the perpetuation of knowledge.

There are two important aspects of a pragmatic approach to truth. The 'conform' and 'transform' theories of truth bring out a distinction between those who consider that truth is a relation between a 'statement' and an 'antecedent reality' and those who believe that truth emerges from a situation of 'uncertainty through the transformation of an unsettled into a settled situation'. Despite debates on these two theories of truth, some argue for conformist while others opt for transformative ideology. The conformist approach regarded objectivity as an inter-temporal relation where the search for truth is conceived as an outcome of a pre-destined situation and has a correspondence between today's truth and that of the pre-destined outcome. It is observed that the debates provoked by a forward-looking, normatively-determined community-based drive towards the truth may throw up many alternative suggestions for sustainability theorists. However, if we find it comfortable within the framework of pragmatism within the 'transform' tradition, which may specify more details regarding the progress we can make towards sustainability, then there is an exciting future for 'Environmental Pragmatism'. But if pragmatists follow the normative nature of logic and enquiry to stress on experimental reassuring to bear upon goals, values and facts, then environmental ethics is likely to become an important sub-field of adaptive management science rather than being an abstract part of the humanities (Norton, 2003).

As regards the pragmatism and sustainability theory, there is a need to develop epistemological communities to build guidance, trust and support for managers to have scientific methods for better conception of sustainable living and sustainable policies. If there are any non-definable sustainable outcomes, these must emerge out of active social experimentation and learning. No doubt, the truth is that which prevails in the long run but both truth and the right to adaptive behaviour need to be understood in an adaptive-evolutionary sense. Further, there is a need to protect bio-diversity as for any extinction of species or destruction of a unique ecosystem, we are under obligation to protect as many as possible particularities of the biological world for future study as well as to act in order to perpetuate the community of inquirers to know more about rain forests, millions of species or natural ecosystems. It is also argued that if our basic commitment is to survive, then within the Darwinian epistemology, the pragmatic analysis based on value, action and science should be complementing pluralism with regard to ethics. Besides, for the survival of our culture and genes, we need to emphasize a new non-instrumental ethic having survival value for sustainable living (ibid.).

If we examine the relevance of environmental philosophical orientations, as briefly outlined above, for the Indian society, it may be argued that radical environmentalism or 'deep ecology' (as it is the nomenclature and distinguished from shallow ecology being pursued by the established environment policy), may not be a plausible solution for environmental sustainability. We have democratic governance and we cannot advocate and pursue regimentation of the principles or policies related to environmental preservation and protection despite the fact that 'deep ecology' is not radical. As a part of civil society, there is a mushrooming of non-governmental organizations (NGOs) which have spearheaded a large number of environmental movements, but there are not many which can acquire a militant-like organization to combat the menace of environmental hazards and degradation of various kinds. More than 200 environmental laws have been enacted but their implementation is inadequate and poor. Moreover, the NGOs' interventions as a part of environmental policies are limited in scale and lack replicability on a larger scale across the country. But, the experimental experience of some of these can be appropriately detailed into different stages of environmental policies and plans for environmental sustainability. In totalitarian societies such as the erstwhile Soviet Union or China, radical environmentalism may prove more effective than in the nations having democratic governance. For example, sanitation campaigns under coercive regime like China became a great success.

As regards eco-feminism as an environmental philosophy, it has a great future for our society. Nearly half of our population is that of females and being an agricultural economy (nearly 70 per cent of our population lives in villages), women are actively involved in agricultural and allied activities. They have the necessary potential to preserve and protect our environment. They only have to be made knowledgeable and skillful for preserving and protecting our plant and animal species. Besides, we must develop the right kind of attitude and values towards environmental cleanliness related to healthy living conditions based on sanitary living, clean water intake and by having air free of pollution. As a philosophy of environment, eco-feminism has a great relevance for our society since it is rooted in the interconnectedness of members of the family. No doubt the Indian family structure is changing, but still, we have a large number of families, especially in villages, which are joint or extended in structure. Women can play a significant role in environmental protection as their mothering skills can go a long way in the healthy growth of children in homes as well as in the community. It has been rightly remarked. 'You give me a good mother, I shall give you a good nation.' This does not imply that men do not have any role in environmental protection. In fact, the roles of both men and women, instead of being confrontal, need to be complementary to one another for building a good society.

For a developing society like ours, 'pragmatism as an environmental philosophy' is highly desirable. One, it is based on seeking the truth on environment through experience or experiment within the normative structure of the community. The 'conformist' or 'transformative' approach pursued is likely to throw a number of possibilities or alternatives for sustainable living. Further, it advocates biodiversity and diverse cultural life as a part of adaptive management science for suitable application in order to find plausible solutions to various facets of environmental sustainability. This approach, in combination with eco-feminism as a part of environmental philosophy in comparison to the tenets of 'deep ecology', may provide better environmental justice and be more ethical for a large population which is poverty struck and living in a state of destitution. The conditions in which these people (poor) live need improvements for their sustainable life and living.

It is well-recognized that in view of fast deterioration and destruction of natural resources as a consequence of intensive and extensive industrialization processes, nature may not be in a position to restore or regenerate itself on its own. As such, intervention by individuals is necessary for environmental preservation and protection. This brings us to the new technological breakthrough on the planet which has both positive and negative consequences for mankind. Largely, these provide gains for a few and pains for the majority. In addition, these cause many kinds of environmental problems that put the majority of population in the developing nations as members of risk societies for healthy living. Besides, these pose serious divisive tendencies in an already highly-stratified social structure in these nations. In the subsequent section, we may examine the implications of such new technologies for the betterment of human groups.

New Technologies, Health Risks and Environmental Hazards

New technologies which have emerged over a period of time are likely to influence social life of people on the planet in the 21st century. In the coming decade or two, these can become major sources of social change. For example, stem cells from cows and fish may produce muscle tissues without growing the complete animal to produce protein foods, thus reducing the need for food and fishing industries.

Further, we may be able to reduce the use of wood pulp-based paper with digital reusable artificial paper. In the next 25 to 50 years, we may be able to reduce pollution and raise the living standards of the poor by reducing input per unit of production and distribution through nanotechnology. The solar power satellites may supply much of the world's electricity without producing green house gases or nuclear waste. Besides, the personalized food increase may enhance mental performance and we may be able to manufacture new forms of life by creating genes and arrange them in unique sequences. Also, rural micro-enterprises may flourish in developing countries with their primary income being through e-commerce.

These and many more new technological inventions and discoveries have the potential of solving many pressing problems and improve the life of the people on this planet, but these also have certain dangers and hidden consequences which may raise serious ethical and moral issues. For example, the low-cost methods to determine the sex of a child are available in China and India (both these nations constitute nearly 40 per cent of world's population and there is a preference for a male child). This may affect the demographic profile of humanity and it may get drastically altered. Further, biotechnology may feed the world, but it may also create serious environmental problems and eliminate the livelihoods for a vast number of people. Even animal cloning may improve quality and cut costs of animal products, but it raises ethical questions about the potentials of human cloning. Research evidence in genomic and stem cell has indicated that human life can be considerably extended. This may to a greater extent affect and alter social and financial commitments. Scientists have been able to determine the minimum number of genes to create life and they believe that they can create new organisms from only chemical ingredients. Some of these new life-forms can be used to have biological weapons, while others can eat toxic wastes and produce fertilizers. As such, there is a need to have public assessment before creating new life forms (Glenn and Gordon, 2002).

The Millennium Project of the American Council for the United Nations University identified and made an assessment of such potentialities of technological development. (This project is a globally decentralized think-tank of more than 1,000 futurists, business planners, scholars, scientists, policy advisers and decision-makers who work for international organizations, governments, corporations, universities and NGOs in over 50 countries). During the last three years, this project has used this network to identify and study issues, opportunities and actions. These have been clubbed into 15 global challenges that the humanity faces at the millennium.

Glenn and Gordon (2002) posit them in a question form and inter alia, these focus on issues related to (1) sustainable development achievement, (2) sufficient provision of clean water to every one without conflict, (3) maintain a balance between population growth and resources, (4) the emergence of genuine democracy away from authoritarian regions, (5) making policy more sensitive to global long-term perspectives, (6) the convergence of information and communication technologies globally for everyone, (7) encourage setting up of ethical market economy to reduce the gap between the rich and the poor, (8) reducing the threat of new and emerging diseases and immune micro-organisms, (9) improve decision-making capacity in the context of changing of the nature of work and institutions, (10) reduction in ethnic conflict, terrorism and use of weapons of mass destruction through instituting shared values and security strategies, (11) changing the status of women to improve human condition, (12) stoppage of organized crime from becoming more powerful and sophisticated globally, (13) have safe and efficient ways for growing energy demand, (14) acceleration of scientific and technological breakthroughs to improve the human condition, and (15) the incorporation of ethical considerations into globally arrived decisions.

These challenges have social-technological dynamics and need examination in the context of new technology such as information and communication, biotechnology, nanotechnology and space technology. These indicate some trends in the globalization of science and technology that need redressal attention of the United Nations.

We need to look at our future, especially with regard to our own species and our relationship with other species and the environment. We may see significant and fundamental changes among the humans over the next few years and decades. These changes will have serious implications, both positive and negative, for our societies in terms of equity, our understanding of ourselves as well as for human diversity and pluralism. It is argued that in a fast-moving future society, everyone would be considered in 'need' of being fixed and enhanced but those who could not get enhanced will essentially be seen as impaired and encounter 'disableism' similar to what disabled people have always experienced. Further, it is pointed out that disableism which is discriminatory, oppressive and abusive behaviour is based on the belief that those who are not performing as compared to those who are performing, as required by the normative pattern of society, are inferior. This may pose a significant human rights challenge in the future and the society in its endeavour must counter these trends to ensure pluralism as a value for human diversity (Wolbring, 2009).

A number of technologies as a byproduct of scientific investigations are emerging and these are contributing towards human performance enhancement. In fact, there is convergence of a number of new technologies such as nanotechnology, biotechnology, information technology, cognitive sciences, synthetic biology, chemistry, material sciences and physics and other science and technology products and processes for human enhancement.

The US National Science Foundation (NSF) Conference in 2001 brought out the concept of NBIC, that is, convergence of nanotechnology (Nano), Biotechnology (Bio), Information (Info) and cognitive and neurosciences (Cogno). The conference highlighted the convergence of these technologies for improving human performance in future for expansion of human cognition and communication, human health and life extension, new tools for national security and counter-terrorism and for new possibilities for learning and education. Further, many different sciences and technologies are likely to converge at the nano-scale and increase our abilities to enhance our bodies and brains in terms of structure, function and capabilities beyond the 'species typical' or what is normal for humans (ibid.).

Before we take up the implications of human enhancement technologies for health, ability-disability divide, body politics, performance, disablement and pluralism, we may briefly understand the concept of nanotechnology. As described, 'nanotechnology' is the collective term for a range of technologies, techniques and processes involving the manipulation of matter at the molecular (groups of atoms) level; systems that typically possess at least one physical dimension in the range of 1–100 nanometres. Such systems may possess entirely new physical and chemical characteristics, resulting in properties that are neither well described by those of a single molecule of the substance, nor by those of the bulk material. This often results in new existing and different characteristics that can generate a vast array of novel products (International Standard Organization, as quoted by Wolbring, 2009).

It is observed that new sciences and technologies with enhancement potential are likely to occur in a number of areas, including medical intervention. Further, the time is not far off when human performance enhancements become part of the mainstream discourse and receive public acceptance. There are many different forms of enhancement, having many different purposes. Each form and purpose of enhancement has its own scale pitches, social consequences, problems and implications. Let us take the case of human health, the enhancement debate would warrant us to make a distinction between therapy and enhancement, but this is not an easy task as many therapies have enhancement capacities and vice-versa. Moreover, many therapeutic interventions, later on, can be used for non-therapeutic purposes. In this context, the meaning of health and disease as concepts become contested and more variations of human body (structure and functioning) are viewed as deviations and diseases. As such, the meaning of concept such as health, disease, disability, impairment, well-being and even medicine assume importance in the enhancement discourse.

Although the WHO definition of health (1946) provides a social model of health which combines 'medical health' and 'social health' under it, yet the policy-makers, over the years, are interpreting health in a narrow sense as 'medical health' or 'medical illness', thus advocating a 'medical model of health' rather than treating it under 'social health' and 'well-being'. Another current conception of health is the trans-humanized model, where health is not to be viewed as the end-point, that is, till someone is considered healthy if his or her biological system functions in a normal manner. It is suggested that despite a person being medically healthy, he or she has some limitations and defectiveness in the health and the new technologies can contribute towards constant improvement. Enhancement medicine is a new field which provides remedy through surgery, pharmaceutical implants and other means. Besides, the increasing capacity of science and technology to help modify the appearance and functioning of biological structures, including human body, is a new kind of ableism. (The trans-humanized form of ableism is based on a set of beliefs, processes and practices that function beyond traditional boundaries of human species.) Trans-humanists advocate the ethical use of technology to expand human capacities to have better minds, better bodies and better lives (Wolbring, 2009).

In view of what has been mentioned above, Wolbring (2009) highlights that human enhancement technologies cannot be stopped by regulation alone. No doubt, enhancement is desired by people because of competitiveness, the fear of losing out and by the ideas of possible right to individual self-fulfilment, yet we need to shift these priorities of people and that of society to focus more on valuing community and diversity. Every new technology results in marginalization of a group of people and creation of new inequalities. As such human enhancement will become an enabling technology for the few but a disability technology for many. This divisiveness in the society will create further inequalities as only those who could afford to seek these human enhancement technologies, may be in a position to send their children to better schools and provide them with personal connections and information technology, which may not be accessible to the less privileged. This may have serious implications of distributive injustice and its impact on social cohesion and global stability. Further, are we prepared to have a society having full freedom and act responsibly to choose and shape our own and our children's abilities and traits? Also, the question of 'person' vis-à-vis a 'non-person' assumes significance as those who cannot avail of enhancement technologies may become as a 'non-person' and a large number of persons will be termed as impaired because they cannot be improved beyond what is normal for our species (the techno poor impaired). This may lead to the danger of decreasing pluralism, tolerance and appreciation of human diversity among humans. There is a legitimization process taking place of the enhancement technologies and this results in commercialization and commodification of abilities in military, gaming industry, especially computer games and sports, etc. This may also shift the emphasis to enhancement drugs rather than to sick people's drugs.

New enhancement technologies, besides curing many kinds of illnesses, will also play a significant role in tackling climate change through 'geo-engineering' and by adapting to climate change by alluring not only the physical and socio-economic environment, but also the human body. It is suggested that if living conditions change due to global warming, we shall have to cater to the demand and marketing of alternation of the human body and mind. As such, we need to discuss the social impact, risk and governance and lack of participation by marginalized and directly-affected groups in the development of science, technology and social policy. Therefore, the scientists, commercial groups, government decision-makers and civil society organizations and people who are at the helm of social movements must get together to thrash out issues related to ability studies, ableism foresight, ableism ethics and ableism governance, for arriving at just and responsible decisions (ibid.).

This perceptible and critical examination of emerging human enhancement technologies clearly brings out that the divide of 'ableism' and 'dis-ableism' and of 'person' and 'non-person' is likely to grow among a large proportion of persons in the developing nations. They are being subjected to health risks of various kinds as most of them are marginalized because of their inabilities to avail enhanced health benefits. As stated earlier, this will not only impair physical and social environment, but would also warrant adjustments of human body not only in the global warming context but also in many other environmental hazards, degradations and crises. This raises a fundamental question about whether or not we shall be able to provide environmental justice to large marginalized persons in the developing nations. Further, we need to address the question of ethics in the context of environmental preservation and protection so as to promote a pluralistic approach by valuing human diversity.

Voices for Environmental Justice and Ethics

Because of the divide between the North and South (developed and developing nations), either of them has started believing in 'blame theory'. The developed nations accuse the developing nations for the state of environmental pollution, degradation and hazards prevailing currently. They want the developing nations to cut on pollution and various types of emissions. On the other hand, the developing nations make a clarion call that the present state of environmental crisis is because of emissions caused by heavy industrialization of the developed nations and they need to be compensated for managing their environmental problems by allowing them to have access to clean technology available with them at a cost which they can afford. This raises the issue of environmental justice to those who are less privileged, poor and living in insanitary conditions, vulnerable to various kinds of diseases.

Environmental justice is basically concerned in providing distributive justice. There are three main reasons for focusing on it. First, we are involved in the environment more than we are aware of it. As such, we have to have greater self-awareness so as to imbibe prudent behaviour towards it. Secondly, distributive justice has not been related to that extent for environment as for other areas of concern. Thirdly, environmental concerns have to watch broader spectrum, that is, relationships not only among people who live in the same society but also among people who live in different societies at the same time; between people of the present and those of the future, between human and non-human animals and the people and the biosphere in general (Wenz, 1996).

Justice assumes significance for people who do not get their fair share in a situation, that is, when they do not get what they want or need, because of the gap between the demand and supply. In a highly divided world, can the unequal partners in power and resources negotiate and communicate effectively for decisions that may help the present and future generations for environmental sustainability? This is a serious question to be addressed by global leaders. No doubt, a number of environmental movements have emerged on the scene globally as well as both within the developed and developing nations to seek a fair share and justice for the poor and less privileged people from the power holders, yet they have to go a long way in achieving their environmental goals for the marginalized populations and affected groups. The grassroots organizations (NGOs, social action groups, people's movements, etc.) have become necessary to seek civil rights and environmental protection, especially in highly stratified societies based on caste, class, race or gender. We are no exception to it, but there is a need for better cooperation and coordination between the governmental and non-governmental organizations for effective results in tackling environmental issues.

Despite 'green peace' movements, which have provided significant improvements in environmental protection over the past several decades, even in the most developed nation like America, millions of people continue to live in unsafe and unhealthy physical environments and are exposed to greater health hazards in their homes, on the jobs and in their neighbourhoods when compared to their more rich counterparts. (These groups are of African American, native Americans on reservations and Mexicans in the border towns, who face economic exploitation, racial oppressions and devaluation of human life and the natural environment).

It is observed that environmental justice beyond US borders needs to be extended to communities and nations that get threatened by the export of hazardous waste, toxic products and 'dirty' industries. In fact, industries and governments, including the military, have always exploited the economic vulnerability of poor communities, states, nations and regions for their unsound and 'risky' operations. Environmental justice warrants that no community or nation (rich or poor) should become a 'sacrifice zone' or dumping ground and should show commitment to protect public health and the environment (Bullard and Johnson, 2009).

What kind of environmental justice can be provided to a nation like ours which is highly stratified on caste and class basis and which has no 'clean technology' developed indigenously because of inadequate Research and Development (less than 1% of our GDP)? This question acquires greater significance in view of the huge financial crunch to meet the basic health needs such as water, sanitary living conditions and a reasonable habitat for a large majority of our population who are marginalized, poor, malnourished and living the life of destitutes (dalits, poor villagers, urban slum dwellers, etc.). We have failed to accomplish the environmental goals of providing pollution-free air, water, housing and sewage facilities to nearly 35 per cent of our people who live below the poverty line, despite enactment of more than 200 environmental protection laws. Are the power holders through the state apparatus insensitive to growing concerns for environmental preservation and protection for the decent living of our present as well as future generations? This raises a moral issue for political masters and administrators who have failed to provide a more just and equitable society to our people. The widening gaps between the affluent rich class and the poor people require immediate actions because justice delayed often becomes synonymous with justice denied.

Even while accepting new technologies such as nanotechnology, biotechnology, information and communication technologies which will provide many positive advantages in the field of medical sciences, agricultural production enhancement and quick information and easy access of network across the world, there are many negative consequences that need to be taken care of. The use of these technologies must be objective and to the extent possible value-neutral. We must prescribe and subscribe to certain standards, norms and values evolved for providing maximum benefits to a large segment of our population and check their misuse or malpractices. This would necessitate evolving a culture of ethical values which should be altruistic rather than mercenary or harmful to the society.

The natural order is considered an order of justice and as such it is to be perceived as a moral order and may be thought equivalent to 'right'. In other words, what is 'right' is moral and what is 'wrong' is immoral. Nature is derived from the Latin word 'natus', which means birth and it is the 'source', the well-spring of life and life is after all an entirely gratuitous gift, owed to no one (Holmes Rolston III, 1979, as quoted by Mathews, 1996). The natural order is not only moral as it secures the long-term good of nature but it is also just and generous (Mathews, 1996).

It is generous in the sense that it rejuvenates and revigorates humans through its process of evolution and adaptation. As such, humans while exploiting the nature to their advantage should show or exhibit certain moral or ethical standards so that the appropriation of nature is not to the benefit of a small segment of the population, but it helps a larger population to make their living at the minimum decent level. In this respect, humans, who have the ingenuity to invent or discover, must ensure that the technology to harness the environment should not be misused or malpracticed, rather it should be employed judiciously so that all could share its benefits in an equitable manner to the extent possible. (For example, the spread of information and communication technologies for human development through 'internet' has operationalized the term 'global' and is viewed as a strong mechanism of change).

Further, it may accelerate the economic development, augment hospital services by tele-medicine, provide better access to the world's knowledge and facilitate self-education and employment. Besides, speed, capacity, miniaturization, speech recognition and synthesis may continue to improve and result in new uses for development and become a global 'conscious' technology. But the increasing dependence on this technology may lead to new vulnerabilities such as fraud, cyber terrorism, information warfare, cultural threats and widening knowledge gaps. Consequently, there may be growth without employment increase, loss of privacy and property rights and emergence of new kinds of problems. The social consequences of all such new kinds of automation, replacing knowledge and judgment, are not well-comprehended and understood. Further, in the context of low-income countries or developing nations, there is a need to assess the likely implication of new automation mechanism as the earlier automation is already affecting physical strength and labour (despite an international digital divide). A 'Computer Economics Survey' expected China to become the second largest Internet-user by 2005, behind the USA. India already has a huge software industry and its exports are growing fast (Glenn and Gordon, 2002).

With regard to the use of biotechnology, there are many remarkable benefits to humans as it provides raw materials for biotechnology industries in the agricultural, pharmaceutical and chemical sectors, but at the same time, there are some worrying concerns regarding the consequences it has for human health and environment. (Biotechnology, which is science-intensive and is based on the scientific knowledge and expertise from disciplines such as botany, genetics, biochemistry and molecular biology, over the last two decades or so has grown in a phenomenal manner and has systematically found new genetic material which has been transformed into new agricultural products, chemicals and drugs. In the agricultural field, there are eight major food sources that have been genetically engineered. These are soya, corn, potatoes, tomatoes, squash, papaya and dairy products (Recombinant Bovine Growth Hormone). In animal husbandry, biotechnology provides animals that help in having production factors for human pharmaceuticals. For example, we have the human gene for insulin and cloning inserted into the germ plasma of cattle and sheep, to have the desired human chemical in urine, blood or milk. The cloning of animals such as cow or sheep helps in making available large quantities of drugs of significant value to millions of people around the world. Further, the mapping of human and plant genomes may promote measures to eliminate diseases having genetic origins or as a result of malfunctioning of genetic material in the body).

There are certain questions being raised about biotechnology having implications for human health through the use of genetically modified (GM) foods to be labeled by the European Union (EU). This may be a difficult proposition as foods are increasingly composed of ingredients having many sources and it may be hard to know whether there are no GM sources in the final food products and how will it affect the health of humans. Even biotechnology has effects on the environment, which are of major significance. For example, genetically engineered traits from one crop can move to its relatives, creating 'super-weeds' which are more difficult to kill. Besides, there are many as yet unknown risks that may become visible to ecosystems. Cloning has remained another major controversial issue in biotechnology. No doubt, it could be of use for growing spare parts for people and as an alternative procreational technique in which a baby's umbilical cord cell can be made use of as replacement cells for adults in the case of bone marrow transplant or brain repair. In addition, stem cells from cows and fish could be used to grow muscle tissue in protein factors, thus reducing the need for cattle and fishing industries. There is an ethical dilemma in this field as a range of stem cells for each of the categories of tissue may be used. For the use of adult stem cells, there is no controversy but for the use of embryonic cells which are more flexible and can be stimulated to become any tissue, there is one. (It is reported by US National Academy of Science Reviews (July 2001) that there were 250,000 unused embryos in the USA which are left over by couples seeking fertility assistance. There are some who consider it immoral to take stem from these embryos because it prevents them from growing into human beings. There are others who believe that these unused embryos will be flushed down into the toilet as the successful parents may no longer need the unused embryos. As such, it is always better to save lives than to destroy them. There is a counter-argument to this, that this may result in a business that may create embryos just to kill them for their stem cells. However, this is refuted by those who are willing to create embryos for stem cell research. We may find new ways of making stem cells without the need to form embryos but interestingly, 'questions of when life begins and who owns it are only now being discussed'). Even nanotechnology through its minimization and precision, no doubt, can help in modifying human genetic and cellular structure besides helping people to build up easily their own wrist watches and computers, yet it is dangerous for creating nanotech weapons, the knowledge of which may be applied by terrorists and trans-national organized crime networks to target individuals and groups. This will destabilize the society and the equilibrium thus disturbed will be a major source of concern for the vitiation of our social environment as well as our physical surroundings (ibid.).

Notwithstanding the above observations, we are already witnessing in our country a decline in sex ratio in some of the states because of easy accessibility of couples to diagnostic tests and equipment to determine the sex of the child. A decline of female child to male in our society, because of the pressure of the social evil like dowry, is causing disequilibrium in our society. (This demographic trend is compelling some of the families, say for example, in states like Haryana and Punjab, where the ratio decline is in favour of females, to seek brides from other distant states for their bridegrooms. This affects the social environment adversely and may result in social disarticulation of families and cohesiveness of the groups). Further, with rapid advances being made in information and communication technologies, computer, mobile phones and other gadgets discarded as waste will become a major source of polluting material for our environment. There is an urgent need to control and monitor these futuristic trends by having certain appropriate regulating bodies to check the violation of our physical, social and cultural environments. This will make our future generations safe against these environmental hazards for comfortable living. Besides, we must evolve ethical values and standards based on love, compassion, mutual aid and friendship, etc., to promote altruism among our young population so that an environmental-friendly perspective is founded on need rather than greed for sustainable living of the majority of people. This will warrant examining the question of environmental sustainability, especially in the context of a developing nation like ours, from the perspective of our rich cultural diversity and availability of biodiversity in abundance, in our natural order.

Culture and Biodiversity: A Boon for Environmental Sustainability

If we want to have a quest for the future with regard to environmental sustainability and development, we must look into our cultural traditions and biosphere which are so diverse in content and form as a part of our cosmos. These define the relationship of our people with nature through a medium of communication based on our shared symbols, idioms and themes. For example,

Vandana Shiva (1992), as quoted in Sheth (1997), rightly remarked that for Asian cultures, the forest remains a teacher and its message is that of interconnectedness and diversity, renewability and sustainability, integrity and pluralism.

The above observation makes it imperative to examine the relevance of our cultural traditions in realizing the goal of a sustainable environmental development.

Cultural Traditions and Biological Diversity

For cultures such as India, Malaysia and Indonesia, the green movement has evolved itself from the forests. Besides, our sages and seers lived in close communion with forests, which formed the major part of nature. Even Kalidasa, Gandhi and Tagore reinterpreted nature as a teacher which provided a model for human society and a harmonious man-nature relationship. Further, India is basically a riparian civilization which flourished along the banks from the Ganga to Cauvery and from Narmada to Brahmaputra. Our cultural seer Tagore rightly observed that the culture of forests has fuelled the culture of Indian society as man for long remained in communion with trees, rivers and lakes, away from crowds. As such, the unifying principle of life in diversity of democratic pluralism formed the basis of Indian civilization (Tagore's play, 'Muktadhara', as quoted in Sheth, 1997).

Sheth (1997) observed that in our traditional philosophical and mythological thinking, every living being is constituted by five basic elements (Panchmahabhuta), that is sky, air, fire, water and soil and all the living world (from grass to forests, rivers, lakes, oceans, hills and the birds and animals) are the natural resources or constituents of them. (This concept of 'Panchmahabhuta' is significant in understanding the ecological balance and well-being on Earth as it provides the organic relationship, interdependence and an exchange on a continued basis between the five elements of the cosmology, which get transformed into one another through a cyclical phenomenon). It is suggested that the world of human beings, the animal world, the flora and fauna, is possible due to the existence of these five elements as they provide life or the 'life-supporting systems' or needs of life throughout our entire span of living by sustaining and promoting our health. It is also reported that nature's constituents such as water, air, trees and plants maintain the spatial balance of Earth while the dawn (Usha) and the night (Ratri) maintain the celestial temporal order. The Indians, in our traditional ethos, are taught to respect various forms of nature such as human beings, wild and aquatic life as well as vegetative life as these are interrelated, interdependent and interwoven with one another. In fact, in Indian traditional thinking, we had a complete holistic perspective of the universe.

It is also observed that we, in India, have applied the cultural lessons at two levels. First, in highlighting relationships between 'society and nature' and secondly, between 'people within the society'. As our society formulated itself on forests and agriculture, we got lessons to socialize ourselves to respect the forms of diversity and as such, we showed concern for every life form, including an insect. Thus, the Indian society is basically eco-friendly and eco-culturally socialized. The other lesson learnt is about renewability and sustainability because it is experienced that sustenance comes from the forest and not from the man-made cities as well as from the fields and not from the factories. It is argued that sustainable reproduction of society is mainly dependent upon the maintenance of diversity in the forests which meets the human needs in diverse ways like that of agriculture, animal husbandry, water and irrigation, housing and health care. We must treat nature (forest) as sacred and maintain its integrity. If its sacredness is violated and the society is deprived or denied of its natural order (nature), then its essential components such as diversity, renewability and sustainability may get lost and the resultant effect will be degradation and destruction of nature (ibid.).

In the wake of the above observations, cultural diversity acquires great significance in preservation and protection of our environment. In an interesting article Jorge Ishizawa (2009) speaks of continued regeneration of biological and cultural diversity observed in Central Andes, covering parts of Peru, Bolivia and Ecuador. He was engaged in Andean Project for Peasant Technologies (PRATEC) with rural communities over a period of two decades since 1986. (The focus of the project was the valorization of traditional agricultural practices and the knowledge implied in them, with the objective of compiling and disseminating technological booklets. More than 2,500 such booklets have been generated and published till date). A few critical issues related to biological and cultural diversity brought out in this project need to be highlighted as they are not only significant but also relevant in encountering environmental problems of various ecosystems.

It is reported that the biological diversity of this region cannot be explained by local geography alone or with its diversity of ecosystems because cultural diversity is represented by a multiplicity of people whose livelihood adapts and get adapted by these ecosystems. Further, biological diversity is not only about wild fauna and flora but constitutes an abundance of cultivated numerous varieties of species that are in existence for millennia. These would not have been possible without sophisticated knowledge of people nurturing such a diversity. (This fact has been well-recognized at the Earth Summit in Rio in 1992 by the Convention on Biological Diversity and Convention to Combat Desertification. Both these conventions have highlighted the value of traditional knowledge underlying the maintenance of biodiversity on the planet).

There are two cosmovisions highlighted, one based on traditional knowledge and the other on Western intellectual scientific thinking. (A cosmovision is concerned in terms of basic ways of seeing, feeling and perceiving the world and gets manifested by the ways in which people act and express themselves). No doubt, modernization of agriculture in this region distracted and even led to partial abandonment of traditional practices in some places, yet elders of the community kept alive the ancient know-how about a great variety of practices such as sowing, planting, harvesting and sorting and reading of natural signs for weather forecasting, including rituals for making a petition for rain, and pleasing the forces of 'hail' and 'frost' and even 'warding off plagues'. On the other hand, the approach of professionals is to apply technical protocol to confirm traditional knowledge validity in line with the criteria of modern science and technology. The outside technical expert or agent not only sets the agenda for the intervention but also the approach to be employed consequently, the local knowledge gets subsumed into techno-science while local approaches get ignored. At the theoretical level (whether, the problem is epistemological in the sense of obtaining knowledge related to local conditions through a techno-scientific approach or it is of ontological nature, that is, our taken for granted understanding of the nature of being and of reality), it is observed that despite modernization of agriculture eroding biological and cultural diversity, it is increasingly being recognized that the communities of nurturers of biodiversity are still alive and active in Peru. This speaks for cultural affirmation of the people of this region. (The concept of cultural affirmation denotes a process by which people living in a place remember and regenerate their traditional practices. In other words, they nurture their own local world and let themselves be nurtured by it) (ibid.).

It needs to be clarified that the concept of culture like that of cosmovision has many meanings and it can be a subject of misunderstanding. In the case of 'Andes' region, it means 'a people's way of life', which includes not only human communities but their deities and the place where they live includes mountains, rivers, lagoons, stones, stars and so on. In this sense, it becomes equivalent to cosmovision and does not refer to a discourse on the customs of people, but is indicative of a form which regenerates its life on a daily basis. The cosmovisions, whether based on traditional cultural knowledge or on Western scientific pursuits, require bridging the two by having respect and affection in the relationship between the two set of communities, that is, scientific and local participants hold the conviction that the conversion between cosmovisions will help in providing certain key insights for healthy regeneration of all involved in the process. Further, the commitment of participants will reflect on 'what is taken for granted' in their respective cosmovision and to look for consequences that affect the healthy regeneration of life on the planet. The dialogue needs a basis of unity. There is a need to revise the role of science as well as have a rethinking on our educational system. The unity in dialogue should be constituted in such a manner that it shows concern for the global mounting ecological crisis

It is rightly remarked that proposals like earth justice or earth jurisprudence need to have a common quest and provide a shift from the anthropocentric world view of the modern Western scientific one to that based on the laws of nature which should subordinate the human law. The initiative for this should come from the Western world and in most cultures, this will get full support. We have now a great diversity of understanding of earth justice in different cultural groups and we need to share such diversity for establishing rich intercultural relations. In this regard, the role of an accompanist is significant as in this process of convergence of two cosmovisions, he or she is also accompanied because every nurturer gets nurtured and has to adapt to the specific circumstances of nurturance. In short, the exchange of lived experience between the accompanist and accompanied has been the main modality that had confirmed affirmation of the cultural diversity and biological diversity in Central Andes, through exchange of visits between human communities by taking part in each other's rituals. These are accompanied by seeds, animals and offerings. The challenge lies in extending this exchange beyond the boundaries of Central Andes (ibid.).

The above observations made in respect of Central Andes culture in the Latin American region confirm the relevance and significance of cultural diversity and with it the biodiversity in preserving and protecting environment. This gets validated an article by Vandana Shiva (2009), 'The Practice of Earth Democracy, Looking Ahead: Experience from 30 Years of Participatory Research and Community Action', which highlights a struggle between a dominant universalizing system of knowledge on the one hand and the plural, resilient and local systems of knowledge on the other, for living in a healthy and sustainable relationship with nature. Before discussing some of the observations made by her, it may be apt to briefly understand the concept of biodiversity.

Biological diversity or biodiversity is a sort of variability observed among living organisms from all sources, including terrestrial ecosystems, marine and other aquatic ecosystems and the ecological complexes of which they form a part. We can conceive diversity at three levels, namely genetic, between species and of ecosystems. The maintenance of biodiversity becomes necessary not only because of its intrinsic value to nature, but it also meets certain critical human needs such as new kind of medicines, food crops, fire and fuel wood and provides essential services such as water purification and carbon storage. As such, it is critical to sustainable development. Recently, we observed erosion of biodiversity at an alarming speed because of human activities (for example, humans utilize nearly 40 per cent of the total primary production on land and we may not be able to recover more than 60 per cent of the world's major fisheries from overfishing without restorative measures. Besides, natural forests are disappearing at an annual rate of some 14 million hectares. Growing population and natural increase in consumption may throw up the challenge of meeting the rising human demands, thus impacting severely the ecosystems).

The main causes attributed to loss of biodiversity are changes in relationships towards nature; the growth in human population and natural resources consumption; impact of global trade; economic systems that do not value the environment and its resources and inequity observed in ownership, management and flow of benefits from both the use and conservation of biological resources. It is observed that biotechnology as an intensive science, (covering a side range of disciplines such as botany genetics, bio-chemistry and molecular biology), has the potentiality to contribute to the conservation and sustainable use of natural resources (Zakri, 2002).

As such, along with scientific advances made in the biotechnological sphere, we must also understand the local system of knowledge as an intrinsic value of natural order for preservation and protection of environment for sustainable development.

In the above backdrop, the observation made by Shiva (2009) needs a critical examination for delineating dominant scientific knowledge and pluralistic resilience of the local folk knowledge in highlighting the significance of cultural as well as biodiversity for encountering environmental problems. She points out that while working with Chipko movement in India, she realized that the powerless persons are so, not because of their ignorance but due to the appropriation of their resources by the powerful. Further, literacy is not a pre-requisite for knowledge and ordinary tribals, peasants and women are biodiversity, seed, soil and water experts as they have tremendous knowledge on ecology based on their experiences. (For example, the women of Chipko movement were aware of the fact that their growing poverty and scarcity of water, fuel and fodder were linked with the profits of logging industry and in order to stop commercial logging, they hugged trees. After a decade of their movement, the Government of India imposed a ban on logging above 1,000 metres in the fragile central Himalayas in 1981). Shiva's efforts to involve the local community resulted in declaring Doon valley, her home area, as a green zone and the limestone mining was stopped. Further, the established forestry, aqua culture and agricultural paradigms got shifted from monoculture to diversity and from commerce to sustenance. (Cases against Neem and Basmati bio-piracy were won after challenging the dominant intellectual property rights. Besides, her involvement in Green Revolution resulted in building the global resistance to genetically engineered crops. No doubt, the Green Revolution came with a promise of prosperity and peace through technological inputs, but it failed because it left nature and society impoverished). It is argued that we cannot expect prosperity to come out of violent technologies that destroy the earth, erode biodiversity, deplete and pollute water and leave peasants indebted and in ruins. This false assumption of prosperity should not become a basis for launching the second Green Revolution based on biotechnology and genetic engineering.

Shiva (2009), in her treatise, pins hope on Gandhi's legacy and philosophy for humanity because corporate globalization may result in extinguishing all freedoms of people by having control of trade, technology and property rights. We may be threatened by the freedom of rivers of flow and organisms to evolve. Further, farmers may not be allowed to save seeds and grow crops and consumers may be denied what they need to eat and even the 'know-how' of food production. As such, the basic freedom elements of all species and ordinary humans may be snatched from them because of free trade practices and globalization processes. In fact, globalization is a sort of colonization and commodification of the resources and processes that provide us life, that is, our biodiversity, our food and our water. Inspired by Gandhi during the last two decades, she visualizes Gandhi's spinning wheel (Charkha) as a symbol of liberation and a means of generating self-reliant livelihoods because it was decentred and labour-generating rather than labour-displacing. She argues that a combination of decentredness, livelihood generation, resource conservation and strengthening of self-reliance is necessary to undo the hold of centralization, livelihood destruction, economic and political dependence that had risen because of the industrialization process under colonialism.

In fact, Gandhi's spinning wheel is a sort of challenge of notions such as absolutism and false universalism, which form the basis of the development of science and technology. Further, it is suggested that the obsolescence and waste are social constructs that have both a political and ecological component. Politically, obsolescence results in getting rid of people's control over their lives and livelihoods by treating productive work as unproductive and retaining people's control over production as a part of notion of progress while ecologically, it destroys the regenerative capacity of nature by substituting her diversity with manufactured uniformity. This is nothing but political ecology of technological development, influenced by narrow and reductionist notions of productivity.

She opines that the struggle between the factory and the spinning wheel continues with the emergence of new technologies. In this regard, she refers to the crisis being engineered for farmers and farming as seeds are getting genetically engineered and patented. In response to it, the seed becomes the 'Charkha' of today. (A national movement, 'Navdanya' was born in 1987 to save seed diversity in farmers' fields because the Green Revolution and even the new biotechnologies along with patent monopolies through intellectual property rights – advocated by WTO, TRIP and Convention on Biological Diversity (CBDI) – are threatening to transform the diversity of life forms into mere raw material for industrial production and endless profits). She expresses that ethically and ecologically unrestrained biotechnology will not only give rise to tools for manipulation, but even patents may offer new tools for monopoly and ownership of products which by their nature are free. This process needs to be reversed by preserving seed which embodies diversity and freedom to stay alive like Gandhi's spinning wheel, which was small but had become an important symbol of freedom. Besides, it became a sign of resistance and creativity in small huts and poor families in our villages (ibid.).

Inspired by Gandhi's spinning wheel efforts as a symbol of freedom, 'Navdanya' (meaning nine seeds) developed into a national programme to save seed diversity for farmers in fields. (These seeds represent India's collective source of food security, denoting a diverse ecological balance at every level, that is, from the ecology of the earth to ecology of our bodies). The efforts of 'Navdanya' programme are to involve the farmers as a conservator of genetic diversity and innovator in the utilization and development of seeds, rather than being robbed by the corporate seed organizations through intellectual property rights protection. The farmers and scientists relate to one another now horizontally rather than vertically, to simultaneously conserve biodiversity as well as to have production of food in order to ensure that farmer's knowledge is strengthened and not stolen. In fact, diversity as a way of thinking and living should go beyond the impoverished monocultures in mind.

We know seed is the first link in the food chain. As 'biza' in Sanskrit, means the source of life, as such, saving seed is our duty while sharing seed is our culture. Because of the dominance of global corporations and global markets, the viable operation of small farmers and small farms get destroyed. The resultant effect, since 1997, is farmers' suicides due to heavy debts faced by them, mainly because of rising cost of production and falling prices for farm commodities, triggered by the policies of trade liberalization and corporate globalization. (The National Crime Records Bureau shows 1,82,936 farmers having committed suicides). There is a need to move away from suicide economy to living economies based on Gandhi's vision of 'Swadeshi' (self-reliance), 'Swaraj' (self-rule), 'Satyagraha' (struggle for truth, the basis of civil disobedience) and 'Sarvodya' (influence for all). Gandhi's legacy has the potential to evolve and renew actions and strategies for freedom suitable to our times and context by creating living economies and living democracies based on 'Swadeshi' and 'Swaraj'.

The living democracy movement (Jan Panchayat) is based on the recognition of the fact that 'we can imbibe our everyday actions with the broadest of visions and the deepest of values' This takes into view the principles of 'earth democracy' upholding that we are all members of the Earth family and our profound identity is our 'earth identity'. As such, we are duty bound to protect all life on Earth. (The declaration of Jan Panchayat was made by 200 villages in June 2000 and since then, this movement has spread to more than 6,000 villages for defending the sovereignty of communities in respect of seed and biodiversity. As a part of the declaration, it was stated that 'cows, buffaloes, goats, sheep, lions, tigers and in fact all animals, birds, plants, trees, precious medicinal plants and manure, water soil, seeds are all our biological resources and we shall not let any outsider exercise any control over them through patents or destroy them through genetic engineering As a community, we shall together be the guardians of our biological heritage. To achieve the goal of Jan Panchayat, Biza Vidyapeeth (school of the seed) was started on the Navdanya farm in collaboration with the Green College (Schumacher College) in the UK and some other scholars.

The learning from nature and biodiversity about Earth citizen is based on involvement in practices that embrace ecological values such as connection, openness, generosity appreciation, partnership, enquiry, dialogue, celebration conservation and protection, which though being devalued in the learning processes are, however, essential for our survival Finally, it is observed that the self is perceived as being at war with nature and society rather than be a part of it. This may result in causation of alienation and violence to the being. To get over that and to restore peace and recovery of our ecological selves, there is a need to re-embed ourselves in the webs of life and social relationships to learn ecological consciousness and education by all the Earth citizens (ibid.).

A perceptive analysis by Shiva clearly brings out that in order to encounter environmental crisis mainly caused on planet

Earth by the minds of monoculture of universalism and scientific progress, we must look towards our culture and biodiversity for environment sustainable development. The universal perspective has caused industrialization, coal and oil burning and nuclear energy potentiality, plastic and chemical industries, congestion of population in urban slums, urbanization, water and air pollution, insanitary living conditions, health hazards, etc. We, as humans on the Earth, must relive our cultural diversities embodying love and compassion for all species, including other animals and plants, rich in biological diversity for our survival as well as for providing a minimum decent living to our future generations by creating environmental sustainability. We need to address this question and come out with some plausible solutions that humans on Earth can throw up by cooperating both globally and locally with each other.

Bequest to the Future Generation for Environmental Sustainability

No doubt, as evidenced above, cultural diversity and with it, biodiversity, especially in the context of bio-massed and agricultural developing nations like ours, is a boon for environmental sustainability. Yet, as humans on this planet, we have some responsibility as rational beings for providing appropriate measures to preserve and protect our environment and natural order for survival and good living of our next generation.

It is now observed that each generation, either thoughtlessly or thoughtfully, leaves a bequest to the generation following it. For example, there are two aspects of bequest that the North Americans of the European descent would leave for the future generation. One pertains to the moral, spiritual and cultural while the second is economic. The colonist, by building 'a city on a hill', set an example of spirituality and righteousness but in modern times, the concern for the future revolves around building wealth, great cities and technological power to maximize economic growth. In respect of spiritual aspect of intergenerational bequest, the role of government and public actions is observed to be minimum. This makes religious bequests mainly a matter of private concern, while the public intervention is significant in empowering governments to accomplish the economic and utilitarian aspects of the bequest.

Economists tend to view substitutability between natural and human-built resources as a key issue in their analysis. It is argued that resource shortage is not going to affect prices because new technologies will lead to development of adequate substitutes for depleted resources. Further, what we owe to the future will be by the principles of fair savings and prudent investments. But, there is a need to differentiate. Our bequest to ensure that material as well as human created capital exists in sufficient quantities because the basic question is not 'what to save' but 'how much to save' (Norton, 2003). Economists also speak of existence 'value', which may in effect reduce concern for the future to be a matter of voluntary altruistic consumer choice in order to invest some part of present income for the benefit of the future. This is indicative of weak sustainability as economic analysis based on value shows only aggregated wealth acquired and maintained over time, thereby reducing obligations across generations to maintain a fair saving rate. It is also argued that we cannot be under any obligation to save particular things that the people in the future would want or need as it is difficult to predict as to what they will need or want. Further, there are no standard economic methods to measure these values. In respect of ecology, how is it possible to determine the value or utility or lack of utility for this or that species? No doubt, biota is useful, but it includes not only plants and animals but also soils and water. We must have a more structured and sensitive biological bequest package to meet a more structured and specific set of obligations by treating weak sustainability as a necessary conceptualization of the intergenerational trust. Besides, there is a need for modifying cost-benefit accounting by applying the safe minimum standard rule of conservation. However, saving a resource is vague in respect of biodiversity because one may have several scales for individuals, populations, species and ecosystems. Besides, can we provide biological or ethical reasons to determine what to 'save' and what to 'restore' (ibid.).

In an ecosystem, the system must generate more biological resources by providing diverse outputs and also by generating and maintaining diversity itself. The diverse biological processes are inclusive in the concept of biodiversity, which includes genetic species within species, the diversity of species and the diversity of habitats in which species live. The policy goal should be to protect biodiversity.

It is observed that diversity creates more diversity and loss of diversity may result in future loss. As such, the self-sustaining and self-creating process is ultimately the source of all biological resources. We need to endeavour to save the resources, provided the costs are bearable. One can, with creative planning and policy through community development efforts, generate short-term economic growth which may be compatible with protecting and augmenting biodiversity. We value, embody and support the varied biological differences because these help in enhancing human options in the future. We do know that people of the future, by having value free choice, will want to have options and opportunities available to them. To be fair to the future generations, our actions today should avoid impoverishing the future and destroying options that may be important in the future. The focus should be to protect biological differences across generations and maintaining freedom of choice. Instead of emphasizing on protecting biological differences in a particular place, we must focus on values and opportunities that the community members in that place are committed to work and protect. Besides, there is a need to identify and protect the differences and processes which are valued by the community habitating that place (ibid.).

In short, we need to identify a socio-cultural process to help communities identify important options with their very sense of the place which they call as home. This will help to build up constitutive values and options arrived at through open discussions involving all stake-holders for a consensus on a set of values that are constitutive of the community and its character. The question that confronts us is whether or not the community members will rationally choose and implement a bequest a trust or legacy – that they will pass on to future generations. We should consider environmental values and valuation while defining sustainability by developing community processes. These have to be democratically governed through the voices of their members, to explore their common values and differences. This may help to save key places and values in order to arrive at the maximum possible consensus as well as for continuing the debate above the differences (ibid.).

What Can be Done?

In respect of a democratically-governed nation like ours, the paradigm for environmental sustainability based on a dialogical and participatory democracy through community discourses and discussions, to arrive at a consensus on environmental issues (that concern all inhabiting the area/region/village/city), appears very promising. This needs to be pursued by political masters, policy planners and administrators who must show bequest to the future generation by shouldering responsibility of providing minimum decent living in the physical, social and cultural environment. What should be done to accomplish this goal of environment preservation and protection? This is a complex, knotty and intricate issue in its accomplishment because of prevailing uncertainties in the environment and lack of commitment and accountability on the part of politicians, planners and administrations at different levels for providing a healthy environment to our society.

Notwithstanding the provision and enactment of laws pertaining to environmental preservation and protection, there is lack of adequate infrastructural and technical services which may salvage environmental problems. These have occurred especially in mega cities and towns due to heavy industrialization, growth of slums within and outside these regions. These places do not have adequate toilet, water and sanitation facilities. In addition, air pollution is caused by heavy density of vehicles emitting dangerous gases, damaging not only the health of human beings but also that of animals and plants. Besides, the solid waste material which acquires a mass because of high-end construction of buildings, roads, development projects like power plants, TV towers, chemical and nuclear plants, weak sewerage system, inadequate facilities for disposal of waste material and refuge, is an emerging environmental problem for a developing nation like ours.

Even the intensification of agriculture through maximization of inputs of chemical fertilizers, pesticides, has caused seepage of these into ground water. Besides, its exploitation beyond its limit level has resulted not only in causing cancer-like disease to rural people, but it has also caused environmental damage such as salinity to soil and desertification of land, thus affecting agricultural produce of different crops. Rural houses are also subjected to environmental hazards because of lack of ventilation facilities and smoke caused by burning of wood fuel and kerosene oil for cooking purposes. The burning of leaves and grain stalks is also responsible for respiratory diseases as well as eye ailments.

We have a huge spread of population across the country in different regions. We have tribals who mostly inhabit the forest areas, deserts of Rajasthan, ravines of Madhya Pradesh, hill regions of North-East, Himachal Pradesh, Uttarakhand and Jammu and Kashmir. They constitute mainly 8 per cent of our population and are mostly dependent on the produce of forests and small farms. Besides, the domestication of animals contributes to their economy. They have a rich cultural heritage of plants which have a medicinal value. Even some of the areas inhabited by them are rich in minerals. However, they are now being exploited by the industrialization processes initiated by wood mafias, manufacturing industries and excavation entrepreneurs because of their vested interests and profit motive induced by market economy in the era of globalization. As a consequence, much of our cultural-based knowledge on preservation and protection of our environment is getting lost or is being stolen.

Besides, we also have nearly 70 per cent of population residing in more than six lakh villages spread in different regions of our country. Despite the onslaught of urbanizing and industrializing influences, the majority of the villages still adhere to centuries old traditions and values as a part of their cultural heritage. In the majority of cases, people practice distilled wisdom based on folk knowledge in their daily social life to remain healthy in an increasingly polluting environment.

Apart from the village-based population, nearly 30 per cent of our people are now living in megacities and large, medium or small towns. Out of these, nearly one-third of them, especially in megacities like Mumbai, Kolkata, Bengaluru, Delhi and Chennai, are living in slums. These are marked by shanty and insanitary living conditions depicting the state of poverty and destination prevailing in urban social life. The majority of population living in these slums is that of migrants from rural areas who come to cities in search of employment, better medical care, better educational facilities, but because of their linkage back home, they still maintain some of their cultural practices, beliefs and values.

This diverse population spread in different regions and areas of the country is an embodiment of cultural diversity based on age, sex, caste and sub-caste affiliations, family and kinship networks and different regions and languages. In each of these cultural groups, there is embeddedness of rich cultural practices and with them biodiversity. These are getting lost due to dominant influences of monoculture mind, causing linear progression of our society based on a set of discoveries and inventions induced by scientific knowledge. But what is desirable in our setting is to promote pluralism, having ramification of our cultural and biodiversified values and practices to encounter our emerging environmental problems. These need to be integrated with the technological-based modern interventions as we lack technical know-how of clean technology. Moreover, we cannot afford the cost of either having or availing such a technology. However, to regenerate our cultural mind along with biodiversity knowledge, we shall have to have a decentralized approach which should open a discourse or dialogue with people in different regions or parts, for their active participation in building up folk knowledge, suitable for sustainable development of our environment.

In this venture, the NGOs (non-governmental organizations) must work in unison with the village panchayats in rural areas and local municipal bodies in urban areas, for making demands on the state machinery for environmental protection and conservation. No doubt, most of the environmental movements have been initiated at the grassroots level in tribal, rural and urban areas but there is a need to establish a networking of these organizations to strengthen their efforts.

Maybe radical environmentalism is the need of the hour as environmental laws to protect the environment have been grossly defied and there is violation of these at all fronts, including occupational hazards in manufacturing industries, chemical industries, power and nuclear plants, etc.

The critical areas for environmental preservation, conservation and protection are a check on population growth and health risks and a spread of awareness or education for all. A check on population growth is a necessary condition for environmental sustainability as a rise in population of humans is depleting the natural resources on this planet. In India, the State took the initiative after the independence of the country, but in due course of time, our efforts in curbing population growth became weak. Over the years, we started putting more emphasis on developmental efforts in Primary, Secondary and Tertiary sectors of the economy, with the hope that the trickle-down effect will take care of the rise in population. As such, we started ignoring the checking of our population growth. Development efforts alone in our society may not reduce population rise unless these are supplemented by educating the couples, especially the needy, the poor and malnourished. We must provide them with family planning services at their doorsteps or next to their doorsteps so that our population gets stabilized at the earliest and becomes conducive for environmental sustainability.

Apart from population check, the most critical factor that may enhance the health risks among humans is the factor of malnourishment. This needs to be addressed along with pollution-free air, water and sanitary living conditions. Nutrition is significant as with nutritious food containing a balanced diet, we can build up adequate resistance among children, women and men to fight against many diseases so as to reduce various health risks. Even the sanitary living conditions, pollution-free air and water are necessary for environmental cleanliness and to avoid contacting communicable disease. Air pollution due to emissions of gases and fuel burning needs to be regulated strictly across the board till alternate energy to propel the vehicles is invented for larger public use. It is a major source of respiratory diseases, which are on the rise in megacities due to traffic congestion. Even supply of potable water for drinking purposes is a major problem in rural areas and in urban slums. Can we supply piped water for drinking purposes to people inhabiting these places so as to check a number of communicable diseases and provide a healthy environment?

To understand and salvage these environmental problems, we need to put them in a sociological perspective. A sociology of environment cannot be envisaged without putting 'people in the centre', as any programme of environmental awareness and education cannot be built without their active involvement and participation. A habitat is important as people live in that environment and they need to be sensitized about the pitfalls that may occur to them if they do not take care of their respective surroundings. There is a need to start environmental awareness programmes in our rural areas and slums as most of the people there are illiterate and poor. As such, the non-formal methods of education should combine a media mix of traditional and modern means of communication to bring home to them the significance of and measures to have a healthy environmental living. Besides, environmental education should become a compulsory part of syllabi at the school, college and university education so as to provide a bequest to the future generation for sustainable environmental development.

Environmental laws will become effective only if there is social awareness about preservation and protection of environment. Even the technical inputs in environmentalism to protect and conserve environment, apart from their feasibility, should be economically viable and socially acceptable. In fact, for environmental sustainable development, we need conducive social and cultural conditions so that economic situation of a large population could be improved since it is essentially a survival need of humans. Besides, we should encourage intermediate technology, which is cost effective and socially acceptable to people for environmental preservation and protection, especially in the rural areas.

In short, the sociology of environment for sustainable development in the context of the Indian society should focus on a multi-dimensional paradigm based on decentralization of the power structure to ensure greater participation of people in debates and counter-debates on environmental issues; pluralism as a way of life to highlight cultural and with it biodiversity; providing intermediate or alternative technology to encounter environmental problems (for smaller scale) and clean technology (for larger scale) issues, strict implementation of laws to check environmental violations and last but not least, through strengthening of population check programmes, measures for reduction of health risks, and education of people regarding environment by formal and non-formal methods as well as media. All that is envisaged in this model needs a political will of our political masters, policy planners and administrators and without their commitment to achieve environmental goals of providing a sustainable development, it may remain a far cry. If they fail in this direction, then the future of our next generation may remain bleak and we may have to pay the price for our negligence.

Above all, in formulating our environmental policy, we should be guided by the principles of 'need' rather than 'greed' and other altruistic values such as concern for others, friendliness, love and compassion for humanity at large and mutual aid or help. These values are intrinsic to our cultural diversity and heritage and need to be pursued in a committed manner as a part of our bequest to the future generation for environmental sustainable development.

References

Bullard, Robert D. and Glenn Johnson (2009), 'Environmental Justice: Grassroots Activism and its Impact on Public Policy Decision Making', in *Environmental Sociology: From Analysis to Action*, Leslie King and Deborah McCarthy (Eds), New York: Rowman & Littlefield Publishers Inc., pp. 63–79.

D'Eaubonne, Francoise (1996), 'The Time for Ecofeminism' (Translated by Ruth Hottel), in *Key Concepts in Critical Theory: Ecology*, Carolyn Merchant (Ed.), Jaipur: Rawat Publications (originally published in the USA by Humanities Press International Inc., 1994), pp. 174–197.

Glenn, Jerome and Theodore Gordon (2002), 'The Future of Technology and the Implication for Society and the UN System', in *Human Development and the Environment Challenge for the*

United Nations in the New Millennium, Hans Van Ginkel, Breudan Barrett, Julius Court and Jerry Velasquez (Eds), Tokyo: United Nations University Press (Indian Edition, Jaipur: Rawat Publications), pp. 110–125.

Guha, Ramachandra (1996), 'Radical Environmentalism: A Third-World Critique', in *Key Concepts in Critical Theory: Ecology*, Carolyn Merchant (Ed.), Jaipur: Rawat Publications (originally published in the USA by Humanities Press International Inc., 1994), pp. 281–289.

Ishizawa, Jorge (2009), 'Affirmation of Cultural Diversity: Learning with the Communities in the Central Andes', in *Development Dialogue* (What Next, Volume II: The Case for Pluralism), No. 52, August, Sweden: Dag Hammarskjold Foundation, pp. 105–139

Mathews, Freya (1996), 'Eco-feminism and Deep Ecology', in *Key Concepts in Critical Theory: Ecology*, Carolyn Merchant (Ed.), Jaipur: Rawat Publications (originally published in the USA by Humanities Press International Inc., 1994), pp. 235–245.

McCarthy, Deborah and Leslie King (2009), 'Introduction: Environmental Problems Require Social Solution', in *Environmental Sociology: From Analysis to Action* (Second Edition), Leslie King and Deborah McCarthy (Eds), New York: Rowman & Littlefield Publishers Inc., pp. 1–22.

Norton, B.G. (2003), *Searching for Sustainability: Interdisciplinary Essays in the Philosophy of Conservation Biology* (Cambridge Studies in Philosophy and Biology), Cambridge: Cambridge University Press.

Rolston III, Holmes (1979), 'Can and Ought We to Follow Nature', *Environmental Ethics*, Vol. 1, No. 1, pp. 7–30, as quoted in Freya Mathews' 'Eco-feminism and Deep Ecology', in *Key Concepts in Critical Theory: Ecology*, Carolyn Merchant (Ed.), Jaipur: Rawat Publications (originally published in the USA by Humanities Press International Inc., 1994), pp. 281–289.

Sheth, Pravin (1997), 'Environment in the Indian Cultural Tradition', *Environmentalism: Politics, Ecology and Development*, Jaipur: Rawat Publications, pp. 15–29.

Shiva, Vandana (1992), 'The Green Movement in Asia', in *Research in Social Movements: Conflicts and Change*, Mathias Finger (Ed.), London: Jai Press, p. 195, as quoted in Pravin Sheth's *Environmentalism: Politics, Ecology and Development*, Jaipur: Rawat Publications, 1997.

——. (1996), 'Development, Ecology and Women', in *Key Concepts in Critical Theory: Ecology*, Carolyn Merchant (Ed.), Jaipur: Rawat Publications (originally published in the USA by Humanities Press International Inc., 1994), pp. 272–280.

——. (2009), 'The Practice of Earth Democracy, Looking Ahead: Experiences from 30 Years of Participatory Research and Community Action', in *Development Dialogue* (What Next, Volume II: The Case for Pluralism), No. 52, August, Sweden: Dag Hammarskjold Foundation, pp. 79–103.

Wenz, Peter (1996), 'The Importance of Environmental Justice', in *Key Concepts in Critical Theory: Ecology*, Carolyn Merchant (Ed.), Jaipur: Rawat Publications (originally published in the USA by Humanities Press International Inc., 1994), pp. 248–253.

Wolbring, Gregor (2009), 'What Next for the Human Species? Human Performance Enhancement, Ableism and Pluralism', in *Development Dialogue* (What Next, Volume II: The Case of Pluralism), No. 52, August, Sweden: Dag Hammarskjold Foundation, pp. 141–161.

Zakri, A. Hamid (2002), 'Global Governance of Biological Diversity', in *Human Development and the Environment: Challenges for the United Nations in the New Millennium*, Hans Van Ginkel, Breudan Barrett, Julius Court and Jerry Velasquez (Eds), Tokyo: United Nations University Press (Indian Edition, Jaipur: Rawat Publications), pp. 264–283.

Index